T0132753

Edited by
EVA MILLESI, HANS WINKLER &
RENATE HENGSBERGER

The Common Hamster (*Cricetus cricetus*): Perspectives on an endangered species

Series editors
Hans Winkler & Tod Stuessy

Austrian Academy
of Sciences Press

OAW

Vienna 2008

Cover: © Claus J. Böswarth

EVA MILLESI, HANS WINKLER & RENATE HENGSBERGER (Eds.): The Common Hamster (*Cricetus cricetus*): Perspectives on an endangered species

ISBN 978-3-7001-6586-6, Biosystematics and Ecology Series No. 25, Austrian Academy of Sciences Press; edited by: HANS WINKLER, Konrad Lorenz-Institute for Ethology, Austrian Academy of Sciences, A-1160 Vienna, Savoyenstraße 1a, Austria and TOD STUESSY, Department of Systematic and Evolutionary Botany, University of Vienna, A-1030 Vienna, Rennweg 14

Contents

Preface A. HERZIG

The Common Hamster is a palaearctic species, widely distributed from western Europe across Russia and Kazakhstan to northwest China. In Europe it ranges from the Netherlands, France and Germany to the west of Russia and south from Slovenia to Bulgaria. In certain areas it is still considered to be a pest on farmland and historically it was also hunted for its fur.

Today, the Common Hamster is among the critically endangered mammals in western European countries and any additional information on its demography is therefore desirable.

While some European countries already have guidelines for the conservation of the hamster, others are still struggling for information on the status of this animal. For this reason, an international meeting to update our knowledge on different aspects of hamster biology is important and fruitful for necessary conservation measures.

The 13th Meeting of the International Hamster Working Group was held in Illmitz, Austria, at the Information Centre of the National Park Neusiedler See-Seewinkel in October 2005.

This volume of the *Biosystematics and Ecology Series* of the Austrian Academy of Sciences contains the *Proceedings* of the meeting, which are composed of two major parts, one offering papers on population monitoring and conservation projects, the other studies on reproduction.

The conservation of the Common Hamster has proved to be complex and difficult not only in terms of field management but also on the policy and jurisdiction level. The laws on nature conservation, e.g. in Germany, only protect the so-called "nesting sites". They do not explicitly cover the habitat of the Common Hamster, even though recent field studies have shown that Common Hamsters use a large area both seasonally and annually.

The protection of populations living in the extreme western part of the distribution range has become a high priority issue. However, monitoring studies on a population in Alsace, France, revealed that a 5-year conservation programme was unable to stop the population decrease.

At least in the western European countries, the current status of the Common Hamster is critical. Hopefully, studies like the ones presented in this volume and forthcoming future activities will brighten the perspectives of this species.

ALOIS HERZIG

Preface E. MILLESI

The Common Hamster (*Cricetus cricetus*) is a fascinating animal, showing an impressive plasticity in annual timing, reproductive performance and hibernation patterns. Nevertheless, habitat destruction and fragmentation have caused a dramatic decline over the last decades. Particularly in the western distribution range of the species only a few isolated populations exist. In some European countries, the Common hamster is currently more or less restricted to strongly altered habitats, mostly with synanthropic character, including agricultural, recreational or even urban areas. Such populations are apparently vulnerable to perturbations due to the high human impact and the scattered distribution patterns.

In this volume we have collected contributions presented at the 13th Annual Meeting of the International Hamster Workgroup in Illmitz, Austria, in 2005. Recent results on monitoring and re-introduction projects as well as on the ecology, behaviour and physiology of the species help broaden our knowledge about the Common hamster and promote international networks to plan, implement, coordinate and evaluate management plans.

We would like to thank all referees for their effort and the following organizations for their support:

- Federal Ministry of Agriculture, Forestry, Environment and Water Management

- City of Vienna, MA22

- Provincial Government of Burgenland

- Provincial Government of Niederösterreich

- Nationalpark Neusiedler See - Seewinkel

- Municipality of Illmitz

- Museum of Natural History

- University of Vienna

EVA MILLESI

on behalf of the Organizing Committee

Population monitoring and conservation projects

The second French Common Hamster (*Cricetus cricetus* L.) conservation program: concept and details

ISABELLE LOSINGER & JANUSCH PÖTER

Abstract: The Common Hamster (*Cricetus cricetus*) population in Alsace, France is severely threatened and declining. The protection of these populations, living in the extreme western part of their distribution range, has become an issue of prime importance for the conservation of the species in Europe. In France, the species acquired the status of a protected species in 1993, and as such became the subject of a first conservation program in 2000–2004. It was implemented by the Office national de la chasse et de la faune sauvage at the request of the Ministry of Ecology and Sustainable Development. The success of the actions that had been implemented was assessed in 2005. The monitoring studies showed that the conservation plan was unable to stop the population decline. At that time, the species was found in only 62 of the 387 Alsatian villages in which it still was present in the early 20th century (WENCEL et al. 2003). Based on these results, it was decided to extend conservation program to the period between 2007 and 2011.

Here, the authors present Common Hamster conservation programs in France and describe experiments conducted during the first national conservation program. Proposals for actions in the future are presented, and all partners concerned are encouraged to contribute to their successful implementation.

1 Introduction

The Common Hamster (*Cricetus cricetus*) is a rodent threatened with extinction. The species experienced a dramatic population decline during recent decades in most parts of Europe, especially in western and central Europe, but the distribution area and population densities have also become reduced in eastern European states (NECHAY 2000).

The Common Hamster may still be found at the western limit of its distribution range in France: today only one small population is left in the Alsace plain, in the Lower-Rhine and Upper-Rhine departments. In the early 20th century, the Common Hamster was present in more than 387 communities (BAUMGART 1996) whereas it only inhabits about 62 today (Fig. 1). The core populations within a radius of about twenty kilometres around Strasbourg (west and southwest of the city) and in a few areas northwest of the Lower-Rhine department are still stable. In the Upper-Rhine department, a small core population is still present along the border of the department. Its decrease is largely due to the inten-

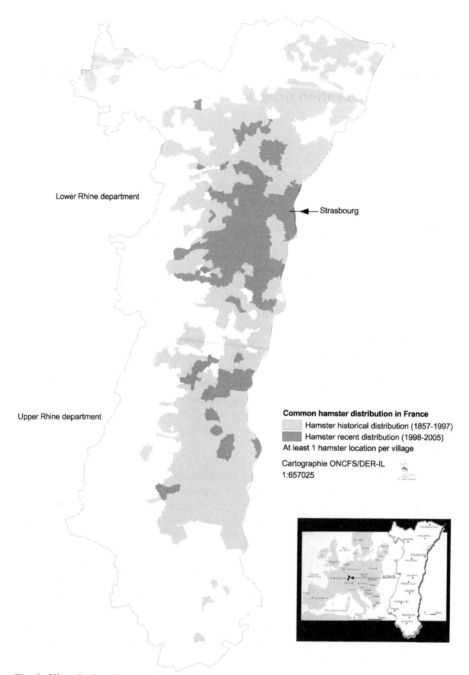

Fig. 1: Historical and recent Common Hamster distribution in France.

sification of agriculture, especially in the second half of the 20th century, and to the changes of agricultural practices in Alsace (LOSINGER & WENCEL 2003) as well as in central and western Europe (BACKBIER et al. 1998). Other factors have also played a role: loss of suitable habitats due to changing land use (new roads, spread of villages and towns), which is often combined with fragmentation of the landscape leading to isolation of the animals and fragmented populations. Confronted with this situation, several measures were taken. At an international level, the Common Hamster is protected by the Bern Convention (1979) and the Council Directive 92-43/CEE on the conservation of natural habitats and of wild flora and fauna. In France, the specific part of this Directive on the Common Hamster was transposed into a ministerial order for the protection of mammals (1981, modified in 1993, 1996, and 2004). The last modification introduced an essential legal element: from now on, not only the species, but also its specific environment are strictly protected. In 1999, based on new knowledge on the biology of this species and Bern Convention recommendations (n°68 and 78 from 04/12/1998) a first specific national conservation plan was formulated and implemented between 2000 and 2004. At the same time Germany, the Netherlands (KREKELS 2000) and Belgium (VALCK et al. 2001) carried out similar programs.

The objective of this first French national program (WENCEL et al. 2002) was to identify the main problems arising from the population decline, and to locally maintain and reinforce the populations southwest of Strasbourg, around Geispolsheim. These measures, which were systematically taken along six complementary axes of operation — i.e., the farmers' acceptance of the species, habitat preservation, population monitoring, public awareness building, reinforcement of the populations, and the development of partnerships with international research teams — should not only stop the present decline of the Common Hamster populations, but also allow for recolonialization of at least some parts of their recently lost territories.

2 Context

In 2004, the first conservation plan for the French Common Hamster populations was completed. A questionnaire was introduced in 2005 to evaluate the actions that had been carried out (KLEIMAN et al. 2000). This was completed by an interview with persons participating in the plan, and by an analysis of the results obtained by the scientific monitoring programme. The main results are the following (LOSINGER 2005):

(1) The information campaigns and the prevention of damage to agricultural crops allowed the farmers to become better informed about the status of the species. They were able to obtain compensation for damages to crops with a high added value (cabbage, beets, vegetables, tobacco, hops). After 61 certified statements that encompassed 187 burrows, 23 animals were translocated, and the damages due to 101 burrows were indemnified with a total of € 8,984.

(2) The population decline was reduced by the reconstitution of favourable habitats (thanks to management conventions and contracts for sustainable agricultural practices that encouraged sowing of alfalfa or winter cereal crops the hamsters are particularly fond of). Ultimately, the total area under contract, i.e. where hamster-friendly management practices were applied, represented 150 ha in 2004. € 152,000 were distributed to 69 farmers in five years. The evaluation showed that the reconstitution of a network of favourable crops with a specific technical plan adapted to the biology of the species ensures fast colonization of the land under agreements, particularly those where alfalfa is cultivated. These crops can be used as "a population source", allowing the Common Hamster to spread and its population density to increase at the communal level.

(3) Population counts allowed us to assess the species' population levels in most of the communities where hamsters were present. To monitor the population changes in Alsace, a method to estimate abundance levels was validated in 2000 (Wencel[1] 2000). Since 1998, 100 communes have been investigated (86 in the Lower-Rhine/14 in the Upper-Rhine area), i.e. a total surface area of 5,000 ha. In 52 communes the densities were relatively low, and hamster densities reached 0.5–1.8 burrows per hectare in ten communes only[2]. The trend analysis showed a steady decline, with a recent estimate of very low population numbers: 500–1,000 individuals distributed into two main, geographically separated core populations over more than 14,700 ha of arable land (about 190,000 in the 1970's, Fig. 2). No genetic exchange is assumed to have taken place between these separate populations. Moreover, when the first plan was implemented, these two core populations were annually censused. In the first site, close to Strasbourg, the number of burrows decreased by 69.2 % (779 burrows in 2000 to 240 in 2005, Fig. 3). In the second site, situated at the border between two Alsatian departments, the population reappeared thanks to a reinforcement program conducted since 2003 (No burrows in 2000, 41 in April 2005).

(4) In addition to a recent plan to communicate information on the hamster, many organisations have initiated information and awareness campaigns: numerous articles with the portrait of the animal, including reports in newspapers and commentaries in local, national and international radio and TV stations, and excursions for school children.

---

[1] Method was based on monitoring of hamster burrows in spring, in favourable fields (alfalfa, winter wheat) only, along line transects 10 m apart.

[2] Number of observations (1998–2005) = 1,121 burrows; proportion of investigated area was 5,000 ha oot of 22,688 ha of favourable fields (22 %).

Fig. 2: Historical distribution and recent abundance of the Common Hamster in France.

Principal core of population (near Strasbourg)

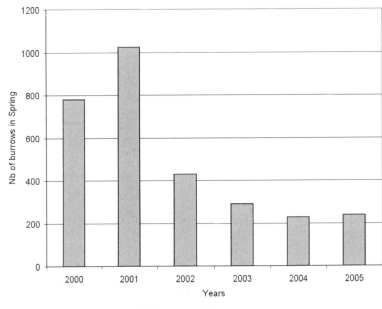

Reinforcement site

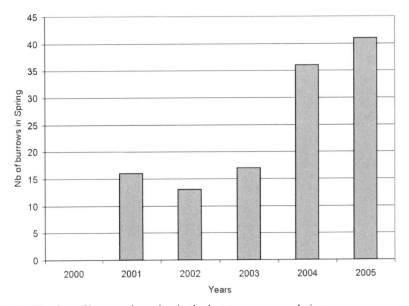

Fig. 3: Number of burrows in spring in the last two core populations.

(5) In late 1999, a hamster farm was set up with animals both caught in the wild and from a farm where they were bred for laboratories. Along the border of the two Alsatian departments, the campaigns to maintain the habitats favourable for the Common Hamster in Alsace proved to be insufficient to preserve the species. Therefore, 175 individuals (75 in 2003, 23 in 2004 and 77 in 2005) have been released within the framework of the population reinforcement program that was applied in this sector since 2003. Two different techniques, as in the Netherlands (MÜSKENS et al. unpublished), were used to release the animals: cage or simple burrows[3]. All animals released were equipped with a transponder so that population changes could be subsequently monitored by capture-mark-recapture (CMR) operations in spring and summer. Twentyone released animals and six "wild" animals were also equipped with telemetric transmitters (TW4 and TW3 Biotrack, UK) in 2004. Telemetric tracking of the released animals revealed a very high mortality rate for individuals bred in captivity: only 12 of the 21 hamsters released into the wild survived more than 78 days. 50 % of those released died in the 11 following days. Predators seem to be the main mortality factor in this rapid and extensive decline (LOSINGER & PETITEAU 2005). Thanks to the CMR operations, 35 new individuals (born in the field) were identified, but none of the released hamsters was recaptured during these operations. This result is similar to that achieved by KAYSER (2003) during a CMR campaign: the author observed no Common Hamsters older than one year and a half. The results reveal a significant problem with respect to the survival of released or captured animals, even if the population is slowly increasing.

(6) A partnership with foreign research teams has allowed an exchange of experience regarding the implementation of the different components of the respective conservation plans. Several professional visits to the countries of the species' home range, along with the annual meetings of the international working group, has yielded new data on population biology, ecology, genetics, and population dynamics.

Moreover, in France, the juridical status of the species has been recently reinforced by a ministerial order of 12/16/2004. The preceding one of 10/10/1996 had granted the Common Hamster the status of a species strictly protected under article 12 of the Fauna-Flora-Habitat Directive. The present population distribution is still to be specified and the breeding and resting sites of the species are presently being defined. This element of "context" totally modifies the strategic

---

[3] Release cages or burrows were only installed in the plots managed under contract, in the presence of the farmers, not far from the areas where burrows had been detected. The cages (1 m x 1 m x 1 m in size) were buried at least 20 cm deep in the ground and a burrow-like, about 30 cm deep hole was dug with the help of a drill to encourage the animal to settle there. The cages were spaced out at distances of at least 30–50 m. Another technique was used to release males, young and old animals: simple burrows were dug and the hamsters were directly released inside them.

approach the next restoration plan should develop: it reinforces the fact that the species should be taken into account in the politics of urban and landscape planning.

Besides, the Permanent Committee of the Bern Convention made recommendations in late 2004 (FERNANDEZ-GALIANO & RIVERA 2005). The Western Europe governments have to:

- Assure the viability of the Common Hamster populations through an active management of populations, connecting isolated reserves with corridors and promoting restocking and re-introduction of the species where necessary.

- Improve the implementation of agro-environmental schemes within the common agricultural policy and target these measures to the farmland areas, in order to ensure the sustainability and minimize the fragmentation of populations.

On these bases, a new restoration plan is being devised for 2006–2010, which could revolve around three axes (LOSINGER & PÖTER 2005):

- In the short term we aim at checking the decline of the population in the peripheral area around Strasbourg and at developing a population core in the population's reinforcement site at the borderline of the two Alsatian departments.

- In the medium term, the connection between the two population cores should be ensured, and the Alsatian human population should be made aware that the species should be preserved.

- In the long term, the survival of a viable Common Hamster population in the Alsace should be realised.

3 The Restoration Plan and what is at stake

The research efforts in the field of population genetics and dynamics conducted in France and by its European partners have allowed the species minimum survival thresholds in the Alsace to be defined (KAYSER 2005; WEINHOLD 2004).

- The genetic effects linked to the populations' consanguinity means that a minimum number of 1,500 hamsters should be present.

- The abundance fluctuations linked to environmental factors mean that the minimum population density should be 4 burrows/ha in spring.

- The populations' fragmentation without any genetic exchanges means that the hamster population should occupy a non-fragmented area of at least 300 ha.

This scientific knowledge forms the basis for further reflections on Common Hamster management, recognizing that the above area sizes and population densities may need adjustment (i.e. an estimated lower population density means a much larger surface area).

To re-establish one or several areas in which a Common Hamster population would be viable, the new restoration plan needs to be based on three priority actions (PÖTER 2005):

(1) Land-use planning that identifies at least two 600 ha areas, termed viability areas (linked to the hard cores of Common Hamster presence) on which all efforts would be concentrated to maximize hamster survival. In these areas the following conditions should be fulfilled: No infrastructures should be present in these areas, nor should any project be planned that involves building an "impassable"[4] infrastructure. In these areas farmers should be commited to cultivating crops that are favourable for the species.

For existing Common Hamster populations, the capacity to colonize their environment should be reinforced by targeted releases of bred animals. This strategy must take into account the carrying capacity of the specific habitat, and in the course of all population reinforcement operations release a sufficient number of animals to ensure their chances of survival. The protocols used to release the Common Hamster and to monitor the reintroduction success will be the same as those used in the reintroduction program (CMR, telemetric tracking).

(2) An adjustment of regulations outside the above-mentioned areas to limit the effects that external projects may have on the Common Hamster population; several recommendations that should be implemented in terms of compensation measures have been defined. The time has come to ensure for the hamster perennially suitable agricultural areas, and follow this up by the implantation of favourable areas.
(3) The acquisition and dissemination of knowledge, notably by initiating an extensive population count and localising specific hamster habitats.

4 Means of action of the restoration plan

4.1 Coordination

The Common Hamster restoration plan shall be steered by the State: the Ministry of Ecology and Sustainable Development and its regional office, the Regional Direction of the Environment in the Alsace. Its actions shall be validated by a steering committee that will meet twice every year. If necessary, several

---

[4] The method to estimate the infrastructurer's barrier effect remains to be defined.

working groups will meet when required to give their opinion on specific topics like public relations or the management of agricultural environments. Each body represented in the steering committee will play a role in implementing the Restoration Plan: the Ministry of Agriculture for habitat management, the associations for the protection of nature to create broad public awareness for the importance of the Plan, and ONCFS for monitoring the populations. No partitioning will be involved: all parties may participate in the actions the others are in charge of.

4.2 The headlines of the conservation plan

In the second conservation plan, all six axes of operation of the first conservation plan, i.e. the acceptation of the species by the farmers, the preservation of their environment, monitoring of the populations and their maintenance *ex situ*, the public awareness campaigns as well as the reinforcement of and partnerships with foreign research teams, are maintained and reinforced. These are suitable tools to carry out the identified priority actions.

4.2.1 Line of operation 1: The farmers' acceptation or tolerance of the species

The Common Hamster damages crops by feeding on them in spring and by gathering food reserves for the winter. The farmers are to be compensated for this. It is now commonly known that the species is protected. In the future, the talks with these farmers should stress what is at stake when a species is threatened with extinction. This calls for developing a full range of actions to prevent such disappearance, to sensitize and educate all segments of the public on this topic, and to present a complete management plan for the species when their numbers are on the rise.

4.2.2 Line of operation 2: Re-creation of favourable habitats

Today, all wildlife species associated with cultivated fields, like the Common Hamster, have considerably declined in Alsace due to the intensification of farming, the expanding road infrastructures, and urbanisation.

Reversing this trend requires reconciling the needs of wildlife and biodiversity on one hand with productive agricultural practices that are also very profitable, on the other hand. At the same time, all potential actors should be sensitised to the application of agricultural practices favourable to small lowland game species.

To avoid habitat fragmentation, it is also imperative to ensure projects that could potentially impact the particular habitats of the Common Hamster conform to the law. The overall objective is to restore a network of uninterrupted habitats that would provide the animals with winter and spring vegetation on which they can feed all year round.

4.2.3 Line of operation 3: Species population monitoring

Counting the population has two benefits: it allows the yearly levels of and changes in the Common Hamster population to be estimated, but also allows the conservation actions for the species to be targeted, and the relevance of the applied protection measures to be evaluated (CAMPBELL et al. 2001). This also involves including the species into urban development planning or road infrastructure projects or assessing the short–term success of the restoration program.

Large-scale counts are therefore essential to determine the status of the reference populations, and to define the strategy that should be applied. The monitoring protocol remains to be defined, but would be based on the monitoring results from the first program. From 2006 on, the entire Common Hamster area (62 villages) will be monitored (see method in Footnote 1). A sampling system (grid mapping of 100 ha) will be carried out each spring.

Moreover, assessing the variations in population numbers is the only way to measure the effectiveness of this second restoration plan. This is why assessing of the success of this population monitoring will remain a priority action of this plan.

4.2.4 Line of operation 4: Public awareness

Most people are uninformed about the Common Hamster. The animal is not popular particularly among the local farmers, although many public relations strategies are available to inform the public. It is the "amiable" aspect of the hamster that makes it a key species for the protection of small lowland game.

To boost local acceptance, every effort should be made to provide insight into the species and its way of life based on scientific experiments. In the long term, this may yield private or industrial joint financing to implement methods to save this species.

4.2.5 Line of operation 5: Ex-situ population reinforcement and conservation

Considering the few Common Hamsters still present in France, it is indispensable that past habitat restoration measures will be completed by population reinforcement. The input of individuals into areas still favourable to the species, or where it has become very rare or absent, should yield core populations that will contribute to the overall restoration of the Alsatian population as its habitat improves.

Therefore, a hamster-breeding program lasting for several years will be implemented. When the operations to reinforce the populations are launched, a management plan for hamster predators will be implemented (if necessary) and, importantly, the release techniques to increase survival rates will be improved.

4.2.6 Line of operation 6: Studies and partnerships

There are many gaps in our knowledge about the biology and ecology of the Common Hamster. In Europe, complementary research is currently being carried out (University of Vienna). Considering out financial resources, the proposal is to discontinue our complementary research effort further and to concentrate on monitoring studies as well as and follow up the research carried out by other foreign teams. Partnerships with the organisations in charge of these studies are thus indispensable for a mutual exchange and to adapt our conservation methods according to the state-of-the-art information.

5 Conclusion

Thanks to the application of specific crop management rotations, the first national conservation plan was successful in reconciling the presence of the Common Hamster and productive agricultural practices. Nonetheless, the population density of the Common Hamster in Alsace still continues to decline. The current population fragments are still below the minimum viable number of 5,000 adults, necessary to retain evolutionary potential and reproductive fitness (FRANKHAM et al. 2002).

Based on the results obtained with this first plan and expert recommendations (International Hamster Workgroup), the second Common Hamster restoration plan will continue the actions launched by the first one and promote its objective, i.e. long-term population viability. Priority items are the restoration of habitat AND corridors for genetic exchange. This would yield a viable population over the long-term whereas regional sites with no opportunities to disperse are doubtful to persist (GODMANN 1998). The goal will be attained by using every avenue to reinforce the restoration plan: e.g. measures to increase the populations, compensation payments in case of damages, improving habitat quality through agreements, mitigation and compensation of future impacts in management programmes, and actions to increase population numbers (conservation breeding and reintroduction).

The long-term viability of the European Common Hamster in Alsace would therefore still be possible if their last core populations could be preserved by applying the current policies for the global conservation of their habitats, and by including all stakeholders: decision makers, farmers, administrations, the local communities, associations and the general public (BISSIX & REES 2000).

The focus on the hamster's historical core populations is a measure of urgency. Note, however, that this wild species must colonize much greater areas to maintain itself over the long term, and without any permanent help. The preservation of viable core populations has become a priority: the only solution to ensure long-term survival is that urban and countryside management projects take the presence of the hamster into account, and that agricultural practices change accordingly.

The work on the Common Hamster conducted since 1996 in France underlined the difficulty of preserving a species, even a strictly protected one, in an intensively managed agricultural ecosystem. The Common Hamster could be considered as an "umbrella" species, and form the symbol of a sustainable, food producing agriculture and of a region suitable for the small fauna of the lowlands. In fact, the Council of Europe recommendation, dated 3 April 1999, qualified the Common Hamster as having symbolic, scientific, ecological, educational, cultural, recreational, and aesthetic values. Future work should not necessarily focus on the Common Hamster as such but on developing of a more global solution that integrates all the small lowland fauna problems (CLARK & HARVEY 2001). The program should support the comeback of all sorts of animals that are currently rare in rural lowlands (European hare, wild rabbit, lapwing, quail, stone curlew, little bustard, skylark), specifically by restoring defined types of rural space.

6 Acknowledgements

This program received funds from the Ministry of Ecology and Sustainable Development, the DIREN Alsace and ONCFS. We would like to thank everyone who contributed to the establishment of this second national plan for the restoration of the Common Hamster populations in the Alsace. We are also very grateful to all the people who made the Conservation program's start and execution a success. Special thanks to EVELYNE TARAN and CATHERINE CARTER, who helped with the English translation, and CLAUDIA FRANCESCHINI who corrected this article.

7 References

BACKBIER, L.A.M., GUBBELS, E.J., SELUGA, K., WEIDLING, A., WEINHOLD, U. & ZIMMERMANN, W. 1998: Internationale Arbeitsgruppe Feldhamster, Stichting Hamsterwerkgroep Limburg; 1998, Der Feldhamster *Cricetus cricetus* (L., 1758), eine stark gefährdete Tierart. — Margraten.

BAUMGART, G. 1996: Le Hamster d'Europe (*Cricetus cricetus*) en Alsace. — Rapport réalisé pour l'Office National de la Chasse. p. 267.

BISSIX, G. & REES, J.A. 2000: Can strategic ecosystem management succeed in multi-agency environments? — Ecological Applications **11**: 570–583.

CAMPBELL, S.P., CLARK, J.A., CRAMPTON, L.H., GUERRY, A.D., HATCH, L.T., HOSSEINI, P.R., LAWLER, J.J. & O'CONNOR, R.J. 2001: An assessment of monitoring efforts in endangered species recovery plans. — Ecological Applications **12**: 674–681.

CLARK, J.A. & HARVEY, E. 2001: Assessing multi-species recovery plans under the endangered species act. — Ecological Applications **12**: 655–662.

FERNANDEZ-GALIANO, E. & RIVERA, E. 2005: Discussion on the implementation of recommendations 68 and 79 of the Bern Convention in favour of the Common Hamster protection. — In LOSINGER, I. (ed.): Hamster biology and ecology, policy and management of hamsters and their biotope. Proceedings of the 12th meeting of the International hamster workgroup Strasbourg (France). pp. 103–107. — ONCFS.

FRANKHAM, R., BALLOU, J.D. & BRISCOE, D.A. 2002: Introduction to conservation genetics. 607 pp. — Cambridge University Press.

GODMANN, O. 1998: Zur Bestandssituation des Feldhamsters (*Cricetus cricetus* L.) im Rhein-Main-Gebiet. — Jb. Nass. Ver. Naturkde. **119**: 93–102.

KAYSER, A. 2003: Survival rates in the Common Hamster. — In MERCELIS, S., KAYSER, A., VERBEYLEN, G. (eds): Proceedings of the 10th Meeting of the International Hamster workgroup, Tongeren, Belgium, The hamster (*Cricetus cricetus* L., 1758): Ecology, policy and management of the hamster and its biotope. — Natuurpunt, Natuurhistorische reeks 2003/2, Belgium: 105–108.

KAYSER, A. 2005: Contemplation about minimum viable population size in Common Hamsters. — In LOSINGER, I. (ed.): Hamster biology and ecology, policy and management of hamsters and their biotope. Proceedings of the 12th meeting of the International hamster workgroup Strasbourg (France). 68 p. — ONCFS.

KLEIMAN, D.G., READING, R.P., MILLER, B.J., CLARK, T.W., SCOTT, M., ROBINSON, J., WALLACE, R.L., CABIN, R.J. & FELLEMAN, F. 2000: Improving the evaluation of conservation programs. — Conservation Biology **14**: 356–365.

KREKELS, R. 2000: Beschermingsplan hamster 2000–2004. Directie Natuurbeheer, Rapport Nr. 41, 60 p. — Wageningen. Ministerie van Landbouw, Natuurbeheer en Visserij 1999: Beschermingsplan hamster 2000–2004. Rapport 41, Informatie- en Kennis Centrum Natuurbeheer, 60 pp. — Wageningen.

LOSINGER, I. 2005: Bilan de la mise en œuvre des activités techniques en 2005 – prévention des dommages aux cultures, suivi des populations de grands hamsters, suivi des conventions de gestion de parcelles culturales, en faveur des populations de grands hamsters et contribution à l'opération de renforcement. 39 p. — Rapport au MEDD.

LOSINGER, I. & PETITEAU, M. 2005: First results of CMR and telemetric study on the released hamster. — In LOSINGER I. (ed.) Proceedings of the 12th meeting of the International hamster workgroup, Strasbourg, France. pp. 53–58. — ONCFS.

LOSINGER, I. & PÖTER, J. 2005: Projet de plan de restauration des populations de Hamsters communs (*Cricetus cricetus*) en Alsace. Période 2006–2010. 160 p.— Rapport au MEDD.

LOSINGER, I. & WENCEL, M.C. 2003: Preservation of Common Hamster (*Cricetus crice-tus*) habitats in France: evaluation of management agreements. — In MERCELIS, S., KAYSER, A., VERBEYLEN, G. (eds): The hamster. — Natuurhistorische reeks **2**: 34–40.

MÜSKENS, G.J.D.M., VAN KATS, R.J.M. & KUITERS, A.T. unpublished: Reintroduction of the Common Hamster, *Cricetus cricetus*, in the Netherlands. Preliminary results. — In NECHAY, G. (ed.): Proceedings of the 11th International Hamster Congress. — Budapest, Hungary.

NECHAY, G. 2000: Status of hamsters *Cricetus cricetus*, *Cricetulus migratorius*, *Mesocricetus newtoni* and other hamster species in Europe. — Nature and environment **106**: 73 pp. — Strasbourg: Council of Europe.

PÖTER, J. 2005: Projet de plan de restauration des populations de Hamsters communs en Alsace Eléments de synthèse. 5 p. — Rapport interne DIREN, Alsace.

VALCK, F., GYSEL, J. & MERCELIS, S. 2001: Soorbeschermingsplan Hamster. 108 p. — De Wielewaal Natuurverniging.

WEINHOLD, U. 2004: City of Strasbourg Southern Bypass Phase II, Piemont-Vosges Expressway and Great Western Bypass. — Expertise report on the Common Hamster. Part I & II. Contractor for DDE, DIREN and ONCFS June/July 2004. 37 p. — Institut für Faunistik.

WENCEL, M.C. 2000: Mise au point d'une méthode indiciaire d'estimation de l'abondance et de suivi des populations de Grand hamster (*Cricetus cricetus*) en Alsace 1996–2000. 22 p. — Rapport interne ONCFS, Gerstheim.

WENCEL, M.C., LOSINGER, I. & MIGOT, P. 2002: Le grand hamster. 66 p. — Publications de l'ONCFS.

WENCEL, M.C., LOSINGER, I. & MIGOT, P. 2003: Evolution de l'aire de répartition du grand hamster au cours du XXème siècle. — Ciconia **27**: fascicule 1, 29–40.

ADRESSES OF THE AUTHORS:
ISABELLE LOSINGER
Office National de la Chasse et de la Faune Sauvage (ONCFS)
Au bord du Rhin - B.P. 15
F-67154 Gerstheim CEDEX
i.losinger@oncfs.gouv.fr

JANUSCH PÖTER
Direction Regionale de L'environnement (DIREN) Alsace
8 Rue Adolphe Seyboth
F-67080 Strasbourg Cedex
janusch.poter@alsace.ecologie.gouv.fr

Movements in translocated Common Hamsters (*Cricetus cricetus*)

Claudia Kupfernagel

Abstract: Between 2002 and 2004, populations of the Common Hamster (*Cricetus cricetus*) were translocated on three different sites in Lower Saxony (Germany). To date, little is known about the success of translocation measures. Hamster mobility has to be considered in choosing the location of compensation areas. In this study, the capture-mark-recapture method yielded information on the remigration capability of Common Hamsters. In total, 72 adult hamsters (31 males and 41 females) were captured, individually marked and transported to their new habitats, the compensation areas. These sites were directly connected to the origin habitats without any barriers. The rate of remigrated animals was 25 % and the frequency was up to 4 times per individual. Male hamsters remigrated a maximum of 460 m, females 260 m. The high tendency towards remigration can be avoided by locating the compensation area far away from the hamsters' familiar landmarks. A univariate regression model predicted that the probability of remigration approximates 0 after 700 m. To minimise the mortality of translocated hamsters, disruptive influences (e.g. roads) should be also located beyond this range. To support population preservation, compensation areas have to be managed in a hamster-friendly manner and must be suitable for the Common Hamster.

1 Introduction

Due to the drastic decline in their population throughout Western Europe, the Common Hamster (*Cricetus cricetus*) is presently an endangered and heavily protected species. A conflict in interest — posed on the one hand by the hamsters' current protection status, and on the other hand by the classification of building sites within the hamsters' habitat — gives rise to disputes. Therefore, the translocation of hamsters is more often considered to be a compensational measure. The conservation and protection of the Common Hamster, requires knowledge about its spatial requirements and mobility. Recent research has therefore been carried out by WEINHOLD (1996, 1998a), WEIDLING (1997), KAYSER (2002), KUPFERNAGEL (2003) and LOSINGER & PETITEAU (2005). Nonetheless, insufficient information and scientific research are available where translocation measures have been introduced. In choosing a suitable compensation area, the minimum distance between the site and the origin habitat (building site) is crucial in ensuring that translocated individuals cannot remigrate.

2 Study sites

The movements of translocated hamsters were investigated at three different sites in Lower Saxony. Two of these study sites were located in Braunschweig (sites I & III), and one near Hildesheim (Fig. 1). These sites were inhabited by hamsters, but were also building sites. For this reason, all resident individuals had to be translocated and were released into compensation areas. The compensation areas (= new habitats) were directly connected with the building sites (= origin habitats) without any barrier (Fig. 2). Throughout the periods of translocation and investigation, the compensation areas and the origin habitats were characterized by the same vegetation cover (Table 1).

Table 1: Characterisation of the study sites.

Study site	Size [ha]	Vegetation	Soil	Date of translocation
Site I	70	wheat, barley	banded para-brown earth	May 2002
Site II	20	wheat, herbs	pseudogley-black earth	May/June 2003
Site III	20	Wheat	para-brown earth	Aug./Sept. 2004

3 Material and Methods

3.1 Study sites

In order to investigate the movements of translocated hamsters, live-traps were activated five days a week on the study sites at the burrow entrances for a period of up to four weeks. Each hamster was marked individually with an ear-tattoo, transported to the compensation area, and released into prepared holes (KUPFERNAGEL 2003). The distances between location of trapping and releasing point varied due to the different sizes of the study sites (Table 1) and the location of the original burrows in the building sites. In order to observe any migrational return to the building site, traps at the original burrows remained activated until translocation measures had been completed. Thereafter, burrow entrances were closed with soil and examined for approximately two weeks and then again before building site preparation. This control ensured that the burrows at the origin habitat were no longer inhabited by hamsters.

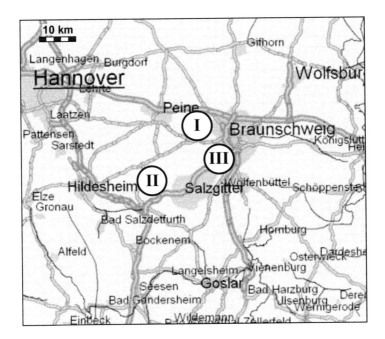

Fig. 1: Study sites I–III in the southeastern part of Lower-Saxony.

Predictions on remigration probability in relation to the distance to the origin are derived based on an univariate regression model. These were calculated from presence-absence data (0 = no remigration, 1 = remigration) and from the distance to the origin using logistic regression (HOSMER & LEMESHOW 1989). Significance tests were conducted using receiver-operation-characteristic-curves (ROC), in which the resulting area under the curve (AUC) describes the discrimination capacity. The AUC-index is tested against the critical AUC-values of discrimination ability, as proposed by PEARCE & FERRIER (2000).

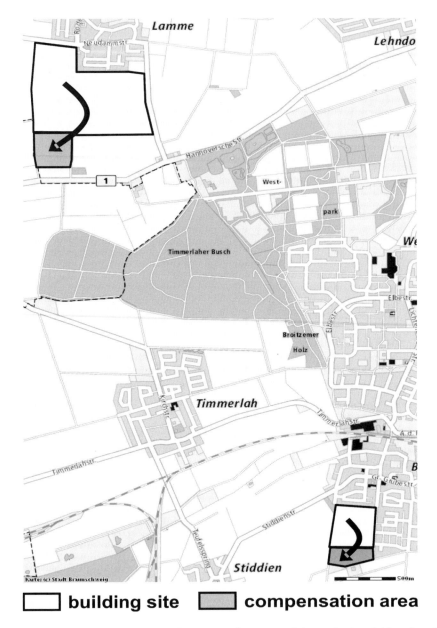

☐ building site ☐ compensation area

Fig. 2: Location of building sites and compensation areas of the study sites I (above) and III (below) in Braunschweig.
arrow = direction of translocation

4 Results

Between 2002 and 2004, 72 adult hamsters (31 males and 41 females) were translocated and individually marked on sites I–III (Table 2). Translocations on sites I and II were conducted in spring, and on site III in autumn (Aug./Sept.). The distance to the origin habitat was between 60 m and 625 m, depending upon the locations of the burrows. The rate of observed remigration was 25 % (18 individuals), and the frequency was 1–4 times per individual (Fig. 3). Remigrated hamsters were able to relocate their origin burrows within one night and they had to be translocated repeatedly. Table 2 indicates that the proportion of remigrating animals increased with decreasing distance to the origin habitat.

The mean distances of movement in returning to the building site were 191 m in females and 220 m in males (Fig. 4). The differences between sexes were not significant, illustrated by the high variation in the distances travelled. The maximum remigration in males was 460 m, in females 260 m. The probability of remigration to the origin habitat decreased with increasing translocation distance and approximated 0 at 700 m and above (Fig. 5).

Table 2: Number of translocated, remigrated hamsters and mean distances to the origin habitat ± standard deviation [m].

Translocated individuals			Remigrated individuals			Distance to origin [m]	
Site (year)	♂	♀	♂	♀	%		
Site I (02)	12	9	1	0	4.76	461.63	96.54
Site II (03)	6	13	2	4	31.58	281.30	45.49
Site III (04)	13	19	4	7	34.38	176.88	74.69
Total	**31**	**41**	**7**	**11**	**25.00**	mean	± SD

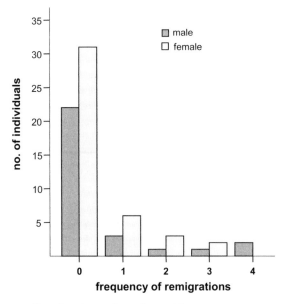

Fig. 3: Frequency of remigrations of translocated hamsters.

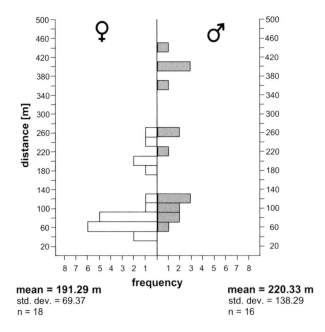

Fig. 4: Moved distances of remigrated individuals after translocation on study sites I-III,
n = frequency of remigrations (analysis includes more than 1 remigration by the
same individual).

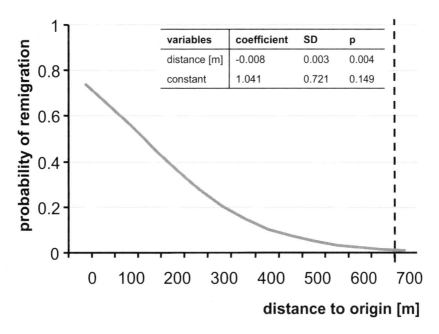

variables	coefficient	SD	p
distance [m]	-0.008	0.003	0.004
constant	1.041	0.721	0.149

Fig. 5: Univariate regression model to predict the probability of remigration in relation to the distance to the origin habitat [m]. R^2 Nagelkerke = 0.22, N = 70, P_{fair} = 0.4265, CCR = 77.19 % (sensitivity: 77.78 %, specificity: 76.67 %), AUC (= 0.848) significantly exceeds AUCcrit. 0.7 at P < 0.05.

5 Discussion

The results illustrate that Common Hamsters can locate their origin habitat after translocation. A remigration rate of 25 %, and moved distances of up to 460 m, demonstrate a comparatively high tendency towards remigration. This high mobility poses a problem, especially if compensation areas are located close to the origin habitat. Such remigration could delay building projects, and expose hamsters to serious predation. Furthermore, the risk of mortality will increase if disruptive factors such as roads are situated within this range of movement. An earlier investigation of translocated hamsters demonstrates that their period of residence at the compensation area is considerably shorter than that of resident individuals (KUPFERNAGEL 2005). Translocated hamsters apparently strive to remigrate.

Under natural conditions the turnover-rate is very high within a hamster-(sub)population. WEINHOLD (1998b) calculated a turnover-rate of 0.87, which indicates that at the end of the active period nearly all hamsters were replaced by new (unmarked) individuals. Thus, the new individuals could be immigrated adult hamsters or juveniles born on the site. This would indicate dispersal of the adult animals. In the Common Hamster, mobility and migration, such as change of home-ranges and burrows, depends on the reproductive period (May until August), (KAYSER 2002). Recent investigations by WEINHOLD (1998a), KAYSER (2001), KUPFERNAGEL (2003) and LOSINGER & PETITEAU (2005) demonstrate that adult males show a higher mobility than females. According to PETZSCH (1950) and WEINHOLD (1998a), a male seeks out burrows of several females until termination of sexual activity in August. Hence, the reproductive exchange is enforced by male hamsters within a population and between subpopulations (WEINHOLD 1997). Females change their burrows after breeding and leave the burrow to their young (KAYSER 2002). The author demonstrates that most changes of burrows occurred in July and that the maximum distance between two successively used burrows was 325 m in both sexes.

What causes remigration and philopatry in translocated individuals? It is debatable whether hamsters know in which direction the origin habitat is located. After translocation the new habitat is first explored (LEICHT 1979, KUPFERNAGEL 2005), initially through random movements and excursions. As soon as hamsters cross familiar landmarks (boundaries of former home-ranges, burrows), they will find a return path. WEINHOLD (1998a) and KAYSER (2002) calculated mean home-range sizes (100 % Minimum Convex Polygon) of 1.66-2.48 ha (male), 0.22–0.44 ha (female). According to WEINHOLD (1996), a hamster can cover at least 300 m per night. In the event of translocation, the compensation area must be located far away from former home-range locations in the origin habitat. Based on the univariate regression model (Fig. 5), the minimum distance from the compensation area (new habitat) to the building site (origin habitat) should be 700 m. This also applies to environmentally disruptive influences (e.g. roads).

Hamsters, however, change and expand home-ranges many times in the course of their active period (KAYSER 2001). In this case, abandoned burrows are preferred, and building sites will probably become repopulated after translocation. To prevent hamsters from settling such sites, they must be transformed into an unappealing environment for the Common Hamster. Migration away from the compensation area has to be minimised through hamster-friendly management. The decisive factor is the level of vegetation cover (KUPFERNAGEL unpubl.), which also reduces the risk of predation. The presence of further individuals in the new habitat is also a factor: during the reproductive season, pre-existing hamsters at the compensation area could enhance site attractiveness. High population densities there, however, could lead to the converse result due to the solitary lifestyle of the Common Hamster. In this case, the site should be rejected. The density at which overpopulation occurs remains to be investigated.

The influence of season on mobility and migration also requires consideration. Hamsters are more mobile during their reproductive period. This might lead to the conclusion that the period following reproductive activity (i.e. from August on) is adequate for translocation measures, due to reduced remigration risk. Note, however, that the portion of remigrated hamsters translocated in August/September (study site III) did not differ considerably from the portion translocated in May under similar circumstances (study site II). Hence, the desire to remigrate and/or the reconnaissance behaviour prevails. Note also that hamsters cease their surface activity from September until October. This makes it essential for translocated individuals to have sufficient time to establish new burrows on the compensation area and to store enough winter stocks for hibernation. Translocation measures should never be conducted while non-independent juveniles inhabit the maternal burrow (according to KAYSER (2002): June until August). Therefore, the optimal date for translocation measures is between the onset of surface activity in spring and the birth of juveniles.

Populations can only be preserved by the appropriate selection and management of compensation areas. A site can only be considered suitable for the Common Hamster if it is a potential hamster-habitat, and this is indicated best by the presence of existing hamsters on the site.

6 Acknowledgements

Special thanks to Prof. Dr. OTTO LARINK, Prof. Dr. GUNNAR REHFELDT and Dr. BERND HOPPE-DOMINIK for supporting this study, ANITA MAURISCHAT and HEIKE BÖHM for field assistance, regional government of Braunschweig and NLÖ for granting required permissions. Parts of this study were financially supported by: Niedersächsische Umweltstiftung and Niedersächsische Lottostiftung BINGO.

7 References

HOSMER, D.W. & LEMESHOW, S. 1989: Applied logistic regression. — New York: Wiley.

KAYSER, A. 2001: Aspekte der Raum- und Baunutzung beim Feldhamster. — Jahrbücher des nassauischen Vereins für Naturkunde **122**: 149–151.

KAYSER, A. 2002: Populationsökologische Studien zum Feldhamster *Cricetus cricetus* (L., 1758) in Sachsen-Anhalt. — Dissertation. Universität Halle-Wittenberg, Germany.

KUPFERNAGEL, C. 2003: Raumnutzung umgesetzter Feldhamster *Cricetus cricetus* (Linnaeus, 1758) auf einer Ausgleichsfläche bei Braunschweig. — Braunschweiger Naturkundliche Schriften **6**: 875–887.

KUPFERNAGEL, C. 2005: Population dynamics of the Common Hamster (*Cricetus cricetus*) on a compensation area near Braunschweig. — In LOSINGER, I. (ed.): The Common Hamster *Cricetus cricetus*, L. 1758. Proceedings of the 12th Hamster-workgroup. pp. 19–21. — ONCFS.

KUPFERNAGEL, C. unpublished: Crop use of the European hamster *Cricetus cricetus* (L., 1758) on a hamster friendly managed study site. — Proceedings of the 11th Hamsterworkgroup. — Budapest, Hungary.

LEICHT, W.H. 1979: Tiere der offenen Kulturlandschaft. Teil 2: Feldhamster, Feldmaus. — Heidelberg: Quelle & Meyer.

LOSINGER, I. & PETITEAU, M. 2005: First results of the reinforcement program monitoring of Common Hamster population in Elsass. — In LOSINGER, I. (ed.): The Common Hamster *Cricetus cricetus*, L. 1758. Proceedings of the 12th Hamsterworkgroup. pp. 53–58. — ONCFS.

PEARCE, P. & FERRIER, S. 2000: Evaluating the predictive performance of habitat models developed using logistic regression. — Ecological Modelling **133**: 225–245.

PETZSCH, H. 1950: Der Hamster. — Neue Brehm-Bücherei. Leipzig, Wittenberg.

WEIDLING, A. 1997: Zur Raumnutzung beim Feldhamster im Nordharzvorland. — Säugetierkd. Inf. **4**: 267–275.

WEINHOLD, U. 1996: Zur räumlichen Organisation des Feldhamsters (*Cricetus cricetus* L.) auf landwirtschaftlichen Flächen in Nordbaden. — Z. f. Säugetierkd. 70. Jahrestagung der DGS, Suppl. **61**: 68–69.

WEINHOLD, U. 1997: Der Feldhamster – ein schützenswerter Schädling? — Natur und Museum **127**: 445–453.

WEINHOLD, U. 1998a: Zur Verbreitung und Ökologie des Feldhamsters (*Cricetus cricetus* L., 1758) in Baden-Württemberg, unter besonderer Berücksichtigung der räumlichen Organisation auf intensiv genutzten landwirtschaftlichen Flächen im Raum Mannheim-Heidelberg. — Dissertation. Universität Heidelberg, Germany.

WEINHOLD, U. 1998b: Abundance of burrows and individuals of the Common Hamster (*Cricetus cricetus*, L. 1758) on intensely used farmland in Northern Badenia. — In STUBBE, M. & STUBBE, A. (eds): Ökologie und Schutz des Feldhamsters. pp. 277-288. — Halle/Saale.

ADDRESS OF THE AUTHOR:

CLAUDIA KUPFERNAGEL
Institute of Zoology
Technical University of Braunschweig
Spielmannstrasse 8
D-38106 Braunschweig, Germany
c.kupfernagel@tu-braunschweig.de

Determinants of above-ground burrow architecture in the Common Hamster

LENKA LISICKÁ, JAN LOSÍK, RADKA KADLČÍKOVÁ & EMIL TKADLEC

Abstract: The Common Hamster is among the critically endangered mammals in western European countries and any additional information on its demography is therefore desirable. Between 2001 and 2004, we explored the relationships between simple descriptors of above-ground burrow architecture – such as the diameter of burrow entrances and the number of burrow entrances – and sex and body mass of individuals in a natural lowland population. We found that the diameter decreases with the increasing number of entrances and increases with hamster body mass. There were no sex differences in diameter or number of burrow entrances. The pattern of seasonal variation in burrow descriptors was weakly developed because of high between-year differences. These results suggest that these simple descriptors of above-ground burrow architecture should not be used to predict changes in demographic structure.

1 Introduction

Recently, there has been much interest in western European countries in studying the demography of declining populations of the Common Hamster (*Cricetus cricetus* L., 1758) (e.g. GRULICH 1980, 1986, 1996; SELUGA & STUBBE 1997; NECHAY 2000; KAYSER et al. 2003; SMULDERS et al. 2003, NEUMANN et al. 2005). The management of small hamster populations is mostly based on population indices such as the number of burrows counted over some defined area (NECHAY 2000). With respect to the general importance of this index, the question arises as to whether the observations on burrows can convey more information than merely the rough estimate of relative density and provide some additional insight into the population structure. This is particularly important in critically endangered small populations in which demographic stochasticity may govern much of the population change.

In this paper, we explore the capacity of simple descriptors of above-ground burrow architecture to predict individual properties of burrow owners, namely their sex and body mass. We therefore examined the relationships between the diameter of burrow entrances and the number of burrow entrances and sex and body mass (age) of individuals captured at these burrows. We also examined the seasonal and between-year variation in burrow descriptors. We carried out the whole study in a central European natural population for which we do not yet have any evidence of decline.

2 Material and Methods

Between 2001 and 2004, we studied burrow architecture of a hamster population in the outskirts of Olomouc city (49°34' N, 17°13' E) in central Moravia, the Czech Republic. The landscape is lowland, about 210 m above sea level, and belongs to the most productive farmland in the Czech Republic. The soil type was a fluvisol with texture characterised as sandy clay to clay loam. The study area of about 30 ha was used for small-scale farming and has a typical mosaic structure with a wide spectrum of crops.

To describe the external architecture of a burrow, we selected two parameters: (1) the number of burrow entrances and (2) the diameter of burrow entrances. Prior to each trapping session, we mapped all hamster burrows in the study area, measured the diameter of each burrow entrance to the nearest 0.5 cm and counted the number of burrow entrances. In total, 60 burrows were described here between 2001 and 2004, of which 77.0 % burrows (n = 46) was used only for 1 season, 20.0 % (n = 12) for 2 seasons, 1.5 % (n = 1) for 3 seasons and 1.5 % (n = 1) for 4 seasons. The repeatedly used burrows were situated on non-ploughed areas. As determinants of burrow architecture, we used sex and body mass (a proxy for age within the sex) of all hamsters captured at the burrow during the single trapping session. Metal mesh live-traps were placed only at entrances which appeared to be active. Each trapped individual was marked using an ear tag (Hauptner & Herberholz, Solingen, Germany) and its sex, body mass, reproductive condition and age was recorded. In total, we captured and marked 296 hamsters.

To analyse the relationships between variables, we fitted several generalized linear models (GLM, procedure GENMOD), assuming that the response variable has either a normal (entrance diameter) or Poisson (number of entrances) error distribution. Variable body mass always entered statistical models in a log-transformed way to meet the assumption of normal error distribution. Seasonal variation was examined by transforming the date to a continuous variable season (with 1 for January 1 and 365 for December 31). The data on burrow entrance diameter from the same burrow are not independent. We therefore applied generalized linear mixed models (GLMM, procedure GLIMMIX with random effects) which accommodate non-independence in data by including random effects (identity of the burrow) in the model structure. We also fitted this class of statistical models to assess between-year variation of the pattern by including a random effect of year. We used the F-statistic as a significance test, with the denominator degree of freedom calculated by the KENWARD-ROGER method. All statistical analyses were done in SAS 9.1.3 (SAS Institute Inc. 2004).

3 Results

The mean burrow diameter was 7.63 (95 % c.i. 5.96–9.30), with values ranging from 5.0 to 9.5 cm (Fig. 1a). The burrow diameter decreased with the increasing number of entrances (GLM: $F_{1,169} = 8.47$, p = 0.004, Fig. 2a). As expected, it grew with body mass of the individuals captured at that burrow (Fig. 2b). However, though the statistical relationship was significant (GLMM, $F_{1,29} = 7.34$, p = 0.01), only a small portion of the variation was explained, with the points being highly scattered around the curve. Even though males in our population were significantly heavier than females ($F_{1,118} = 19.4$, p < 0.001, Fig. 2c), this body mass difference did not translate into different burrow entrance diameters ($F_{1,10} = 1.50$, p = 0.23, Fig. 2d).

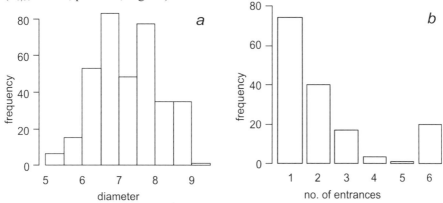

Fig. 1: The frequency distribution in the number of burrow entrances (a) and the burrow diameter (b) in the Common Hamster.

GLM predicted decreasing diameters over the season ($F_{1,334} = 5.48$, p = 0.020). However, if we fit GLMM incorporating year and burrow identity as random effects, the decrease was no longer significant ($F_{1,54} = 2.91$, p = 0.094, Fig. 3a). This suggests that between-year variation in the seasonal pattern cannot be neglected.

On average, a hamster burrow had 1.66 burrow entrances (95 % c.i. 1.09–2.53), the number ranging from 1 to 6 (Fig. 1b). The number of entrances was independent of sex (males: mean = 1.71, 95 % c.i. 1.15–2.54; females: mean = 1.62, c.i. 1.08–2.43; $F_{1,154} = 0.18$, p = 0.67) and body mass ($F_{1,154} = 0.09$, p = 0.77). There was no interaction between these two variables (sex*body mass: $F_{1,152} = 0.12$, p = 0.73). The number of entrances appeared to vary with season in a quadratic fashion (GLM, season*season: $F_{1,144} = 4.62$, p = 0.033), increasing from April to July and decreasing from August to September (Fig. 3b). However, this quadratic trend is no longer significant in GLMM where year was included as a random effect ($F_{1,144} = 2.70$, p = 0.103). This suggests again that the seasonal pattern of variation in the number of entrances varies greatly among years.

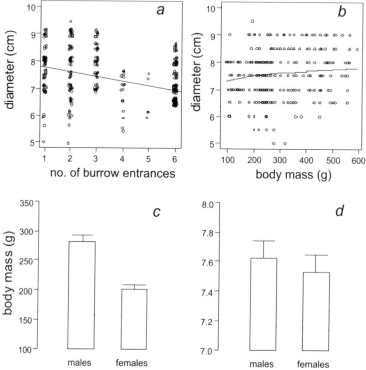

Fig. 2: (a) The relationship between burrow diameter (cm) and number of burrow entrances in the Common Hamster. We used a graphic function "jitter" to show the density of points on the plot. (b) The diameter of burrow entrances (cm) regressed on body mass (g). (c) The difference in body mass (g) between males and females. The bars represent means with 95 % confidence intervals. (d) Differences in entrance diameters (cm) between males and females with 95 % confidence intervals.

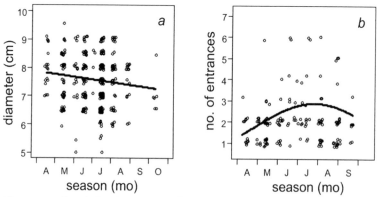

Fig. 3: The seasonal variation in entrance diameters (cm) (a) and number of entrances (b) in the Common Hamster. The graphic function "jitter" was used to show the density of points on the plot.

4 Discussion

Any additional information on the demography of critically endangered western European populations of the Common Hamster that can be derived from the above-ground burrow architecture is highly desirable. Here, we explored the relationships between simple burrow descriptors, such as the diameter of burrow entrances and the number of burrow entrances, and individual characteristics, such as sex and body mass. Entrance diameter decreases with the increasing number of entrances and increases with hamster body mass. There were no sex differences in diameter or number of burrow entrances. Seasonal variation in burrow descriptors was weak due to high between-year differences in the pattern. These results suggest that simple descriptors of above-ground burrow architecture are poorly suited for predicting changes in demographic structure and, consequently, other approaches have to be sought if any additional valuable information for hamster management is to be derived directly from the above-ground burrow architecture.

We found only weak relationships between burrow descriptors and individual traits. The heavier hamsters had the larger entrance diameters but the variation was too high to meaningfully predict mass (or age) population structure. No other study has rigorously dealt with this problem. Only few reports on burrow architecture provide data on the diameter, giving the general range from 6 to 8 cm (EISENTRAUT 1928; GRULICH 1981). Usually, an 8 cm diameter is ascribed to adults, 6 cm to young (EISENTRAUT 1928; NIETHAMMER 1982). GRULICH (1981) observed on average 2.2 entrances per burrow, with the range of 1–12 entrances, which is much wider than that in our population.

That body mass is a poor predictor of a burrow owner primarily reflects the fact that one burrow system is used consecutively over the breeding season by several individuals differing greatly in their body mass. In clayey soils, the burrow can persist for seven years (GRULICH 1981), being re-built regularly after each hibernation with only some of the galleries renewed. Even at one time, a female can be visited by several adult males, and mothers usually share the burrow with young. The decreasing diameter in burrows with many entrances indicates that the galleries dug later have smaller diameters. This may be because older entrances become naturally larger when used by individuals for a longer time. A second explanation is that burrows are extended mostly through the activity of adult females or young individuals building new, often perpendicular, galleries from below, with smaller entrances (GRULICH 1981).

There were no sex differences in the diameter or number of burrow entrances. Again, it is extremely difficult to assign each burrow its owner because of high "traffic" at one burrow system and high seasonal turnover of owners. This could obscure the emerging pattern that females inhabit burrows with more entrances, whose diameters are on average smaller, than do males. Some earlier students of hamster burrow architecture attempted to distinguish between the burrows of different sex by describing the below-ground differences (e.g. EISENTRAUT 1928; KARASEVA & SHILAYEVA 1965). For instance, old females with their young in summer were observed to build the most complicated burrow systems, whereas old males and young animals had the simplest ones. In winter, however, old males more often inhabited more complex burrows (KARASEVA & SHILAYEVA 1965), demonstrating again that the relationship between burrow architecture and gender is very dynamic. On the hand, some researchers questioned the advisability of such an approach (e.g., GRULICH 1981). All these accounts indicate that the above-ground criteria usually correspond little to the underground ones (e.g., RESETARITZ et al. 2005). Consequently, we conclude that no reliable classification of burrows according to the sex of individuals can be achieved using above-ground criteria. Burrows with many entrances do not necessarily belong to breeding females and vice-versa (GRULICH 1981). This calls for other approaches to obtain information on the burrow owner. This is particularly true in areas with heavy soil types like in our study plot, which imposes lower variation in burrow descriptors (cf. GRULICH 1981; RESETARITZ et al. 2005 reporting much higher ranges for entrance numbers).

We showed that the use of external burrow architecture is an unreliable way to obtain additional insight into the demographic structure of a hamster population. Moreover, it is unclear whether simple classification approaches based on inspection of burrow exteriors and readily available for management of hamster populations are valid. Rather, traditional live-trapping methods probably have no simple substitute and remain the primary method whenever detailed information on hamster demography is necessary.

5 Acknowledgements

We thank JAN ZEJDA and MARTA HEROLDOVÁ for their valuable comments on the earlier version of the manuscript. The research was supported by grant GA CR No. 206/04/2003.

6 References

EISENTRAUT, M. 1928: Über die Baue und den Winterschlaf des Hamsters (*Cricetus cricetus* L.). — Z. Säugetierkunde **3**: 172–210.

GRULICH, I. 1980: Populationsdichte des Hamsters (*Cricetus cricetus*, Mamm.). — Acta Sc. Nat. Brno **14**: 1–44.

GRULICH, I. 1981: Die Baue des Hamsters (*Cricetus cricetus*, Rodentia, Mammalia). — Folia Zool. **30**: 99–116.

GRULICH, I. 1986: The reproduction of *Cricetus cricetus* (Rodentia) in Czechoslovakia. — Acta Sc. Nat. Brno **20**: 1–44.

GRULICH, I. 1996: Der gegenwärtige Stand der Hamsterverbreitung (*Cricetus cricetus*) in Tschechien und Slowakien. — Säugetierkd. Inf. **20**: 145–154.

KARASEVA, E.V. & SHILAYEVA, L.M. 1965: The structure of hamster burrows in relation to its age and the season. — Byull. M. O-va. Isp. Prirody, Otd. Biologii **70**: 30–39.

KAYSER, A., WEINHOLD, U. & STUBBE, M. 2003: Mortality factors of the Common Hamster *Cricetus cricetus* at two sites in Germany. — Acta Theriol. **48**: 47–57.

NECHAY, G. 2000: Status of Hamsters: *Cricetus cricetus, Cricetus migratorius, Mesocricetus newtoni* and other hamster species in Europe. — Nature and Environment Series **106**, Strasbourg: Council of Europe.

NEUMANN, K., MICHAUY, J.R., MAAK, S., JANSMAN, H.A.H., KAYSER, A., MUNDT G. & GATTERMANN, R. 2005: Genetic spatial structure of European Common Hamsters (*Cricetus cricetus*) – a result of repeated range expansion and demographic bottlenecks. — Mol. Ecol. **14**: 1473–1483.

NIETHAMMER, J. 1982: *Cricetus cricetus* (Linnaeus, 1758) – Hamster (Feldhamster). — In Niethammer, J., Krapp, F. (eds): Handbuch der Säugetiere Europas, Bd. 2/1 Rodentia II, pp. 397–418. Wiesbaden, Akademische Verlagsgesellschaft.

RESETARITZ, A., MAMMEN, K. & MAMMEN, U. 2005: Structure of hamster burrows and the feasibility of burrow categorization. — In 13th Meeting of the International hamsterworkgroup (October 14–17, 2005, Illmitz/Vienna, Austria), pp. 13. Neusiedler See-Seewinkel: Nationalpark.

SELUGA, K. & STUBBE, M. 1997: Zur Bestandssituation des Feldhamster (*Cricetus cricetus* L.) in Ostdeutschland. — Säugetierkd. Inf. **21**: 257–266.

SMULDERS, M.J.M., SNOEK, L.B., BOOY, G. & VOSMAN, B. 2003: Complete loss of MHC genetic diversity in the Common Hamster (*Cricetus cricetus*) population in The Netherlands. Consequences for conservation strategies. — Conserv. Genet. **4**: 441-451.

ADDRESSES OF THE AUTHORS:

LENKA LISICKÁ

Department of Ecology and Environmental Sciences, třída Svobody 26, Palacky University, Olomouc 771 46, Czech Republic

JAN LOSÍK

Department of Ecology and Environmental Sciences, třída Svobody 26, Palacky University, Olomouc 771 46, Czech Republic

RADKA KADLČÍKOVÁ

Department of Ecology and Environmental Sciences, třída Svobody 26, Palacky University, Olomouc 771 46, Czech Republic

EMIL TKADLEC[1, 2]

[1] Department of Ecology and Environmental Sciences, třída Svobody 26, Palacky University, Olomouc 771 46, Czech Republic

[2] Institute of Vertebrate Biology, Academy of Sciences of the Czech Republic, 675 02 Studenec 122, Czech Republic

LISICKÁ, LOSÍK, KADLČÍKOVÁ & TKADLEC

AUTHOR FOR CORRESPONDENCE:
EMIL TKADLEC
Professor of Ecology
Department of Ecology and Environmental Sciences
třída Svobody 26, Palacky University
Olomouc 771 46
Czech Republic
fax: 00420 585225737
tkadlec@prfnw.upol.cz

Population development and life expectancy in Common Hamsters

CLAUDIA FRANCESCHINI-ZINK & EVA MILLESI

Abstract: In this study we tried to determine life expectancy in a high density population of free-ranging Common Hamsters (*Cricetus cricetus*) in southern Vienna. Sex and age differences in local survival were analysed based on capture-mark-recapture data. In addition, population fluctuations during the active season could be calculated. Maximum life span differed significantly between the sexes. Females lived for up to 2.4 years, males for 2.2 years. Common Hamsters average life spans were 14 months for females and 11.5 months for males. Population fluctuation during the active season was high in both sexes, indicating considerable exchange with neighbouring populations. Winter mortality was similar in juvenile and adult hamsters. High adult mortality rates might be related to reproductive activity which persists over the main part of the active season in this species.

1 Introduction

Common Hamsters are typical steppe inhabitants and are distributed from Russia to Europe (WOLLNIK & SCHMIDT 1995). In Austria, hamsters are abundant in the northeastern part of the country (Vienna, Lower-Austria and Burgenland), where they inhabit different types of habitats (agricultural and steppe-like sites). In Vienna, Common Hamsters have adapted to urban features and live in close vicinity to humans. They can be observed in cemeteries, parks, industrial areas, or green areas of apartment complexes (LENDERS & PELZERS 1985; SPITZENBERGER 1999; FRANCESCHINI & MILLESI 2001). However, this species experienced a dramatic population decline during recent decades mainly in its western distribution resulting in only a few isolated populations in western Germany, France, Belgium and the Netherlands (NEUMANN & JANSMAN 2004). As a consequence the species is now on the Red List of Endangered Species (BAUER 1989) and protected under the EU habitat directive.

Several authors investigated life expectancy in Common Hamsters under different conditions. Whereas VOHRALIK (1975) found a maximum life span of four years in captivity, there is some evidence that hamster populations reach a maximum age of 2.5 years (SAMOSH 1972) in the field. However, in most cases hamsters only live for one season (WEIDLING & STUBBE 1997). Recent studies indicate a maximum life span of three years and significant sex differences were found, with females living longer than males (KAYSER 2003). Although several studies focused on ecological aspects (e.g. SELUGA 1996; WEIDLING 1997), information about longevity and population development in this species, especially in human shaped environments is rare. Several factors might positively affect survival in urban habitats. Predation risk might be reduced due to lower predator abundance compared to more natural habitats, and resources are probably more abundant in the vicinity of humans. There, animals may profit from garbage, or may even be fed. On the other hand, urban populations are highly sensitive to other anthropogenic impacts. Dispersal opportunities are probably limited, and moving over greater distances is dangerous (e.g. crossing streets). This could lead to reduced genetic variation, which makes urban populations highly vulnerable to even small changes within the habitat.

In our study, we focused on investigating characteristics of a population of Common Hamsters living in close association with humans in an area located in southern Vienna, Austria. The aim of our survey was to examine sex and age differences in local survival in this species in an urban environment. To estimate population fluctuations we also focused on turnover rates during the active season as well as during winter. Finally, we aimed at determining over-winter survival and compared it between different sex and age groups.

2 Material and Methods

The study was carried out in southern Vienna from 2001 until 2005. In 2001, basic information about demographic aspects was collected. From 2003 to 2005 intensive studies on the population were done. The size of the study site ranged between 1.2 ha and 4.6 ha. Data collection methods were similar in all years. Due to low vegetation height and the spatially limited study site, many of the marked hamsters could be followed. Hamsters were captured in weekly intervals with Tomahawk live traps baited with peanut butter. Individuals were subcutaneously marked by injecting a transponder chip (PIT tag, Data Mars) for permanent identification at first capture. Hamsters were handled within a black cotton bag (Franceschini & Millesi 2001) and released a few minutes after capture at the place of capture. For field recognition, each animal was individually fur-marked. Data collection was carried out on 5 days per week from spring emergence until the immergence into hibernation in autumn.

Hamsters were categorized in two age classes: juveniles (before first hibernation) and older individuals (after first hibernation).

We calculated survival during the active season (from emergence to immergence) and over-winter survival (from immergence to emergence). Age could be exactly determined when individuals were captured as juveniles. In other cases, minimum age was determined by assuming that the individual was born in August the previous year, the latest possible birth date.

The presence of individuals in the study area was determined by captures throughout the active season. To investigate winter mortality, only individuals that had been present in the study site shortly before the onset of the hibernation season were included. At this period, animals are known to remain resident in the area until the following spring (BACKBIER et al. 1998). In spring, frequent observations of burrows and recapture attempts were made to determine over-winter survival in these individuals. Based on this information, the percentage of winter mortality rates was calculated. Spring emergence date was defined as the first observation of an individual above ground in spring. Recapture attempts were made in the surrounding areas in 4-week intervals to estimate the dispersal distances.

2.1 Turnover rate

This index gives information on changes in population composition. Referring to MÜHLENBERG (1993), changes within the studied population were calculated:

$$T = \frac{I + E}{S_1 + S_2}$$

I =   Number of individuals appearing between capture session 1 and capture session 2

E =   Number of individuals that disappeared between capture session 1 and capture session 2

S1 = Number of individuals trapped in capture session 1

S2 = Number of individuals trapped in capture session 2

2.2 Statistics

SHAPIRO-WILK tests were used to determine if the data were normally distributed. In case of normally distributed data, t tests were used for two sample comparisons, otherwise Mann-Whitney U tests were carried out.

3 Results

3.1 Population density

The number of adult individuals/ha ranged between 19 individuals/ha in 2003 and 9 individuals/ha in 2004. In 2005, 17 individuals/ha could be trapped in the study area. From 2001 on, the population density decreased reaching a minimum value in 2004. This decline was found in both sexes (Fig. 1). Between 2004 and 2005 the number of individuals per ha increased again to 17 individuals.

The sex ratio changed between study years. During the year 2001 more females than males could be trapped. The situation changed in 2003 and 2005, where the sex ratio in marked adult hamsters was slightly male-biased. In 2004 we captured almost the same numbers of females and males.

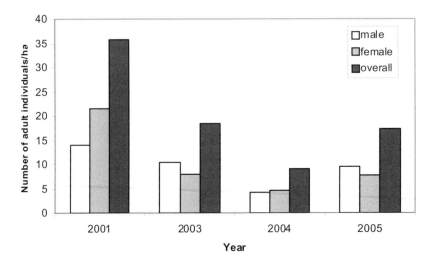

Fig. 1: Population density changes (adult individuals per ha) from 2001 until 2005, inter-sex and inter-year comparisons are shown.

3.2 Winter mortality

Overall, over-winter mortality was 62.5 % in both winters (2003/04: 64.6 %, 2004/05: 60.4 %). From 2003-2004, mortality rates were slightly higher than in the winter before. After both winter periods most marked individuals did not emerge from hibernation (Figs 2 and 3). Winter survival rates in adult females were similar in both years. During the winter 2004/05 adult males had higher survival rates than in the winter before.

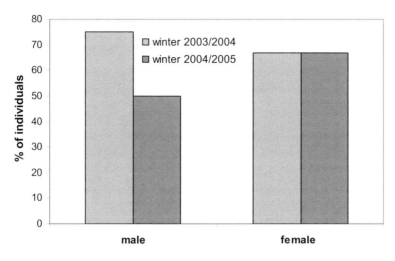

Fig. 2: Percentage of adult individuals that disappeared during the winter periods 2003/2004 and 2004/2005.

Juvenile winter mortality was similar in females in both years, whereas mortality in juvenile males changed: More male juveniles disappeared from autumnal immergence in 2004 until spring emergence in 2005 compared to the previous winter.

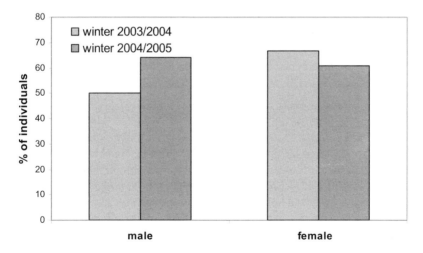

Fig. 3: Percentage of juvenile individuals that disappeared during the winter periods 2003/2004 and 2004/2005.

In general, juvenile winter mortality (60.38 %) and adult winter mortality (64.59 %) were similar. Females seemed to experience lower survival rates (juveniles: 63.77 %; adults: 66.67 %) during winter in both age groups than males (juveniles: 57.00 %; adults: 62.5 %).

3.3 Turnover rates

We estimated the turnover in population density within the study site (Table 1). Seasonal turnover rates (from spring until autumn of the same year) were higher in both sexes and both seasons (2003 and 2004) compared to the over-winter turnover rates (autumn until spring). Males showed slightly higher turnover rates in the course of the season in both years than females. During winter, turnover rates were similar between the sexes and the years studied.

Table 1: Seasonal and over winter turnover rates in both sexes in different time periods.

	Males		Females	
	2003	**2004**	**2003**	**2004**
Seasonal turnover	0.96	1.00	0.84	0.81
	2003/04	**2004/05**	**2003/04**	**2004/05**
Over winter turnover	0.54	0.49	0.47	0.54

3.4 Local survival

Local survival was determined as the time period over which an individual could be observed and trapped in the study area. Based on these data minimum life spans could be calculated.

We found a significant sex difference in the minimum age: Males reached a lower minimum age compared to females (Fig. 4). Whereas females reached on average a minimum age of 14.0 months, males survived over a shorter period of time (11.5 months). When only including individuals that could be followed from juvenile age on, the difference in minimum age is similar (females: 13.2 ± 4.15 SD, n = 14; males: 11.4 ± 4.60 SD, n = 14; p = 0.041).

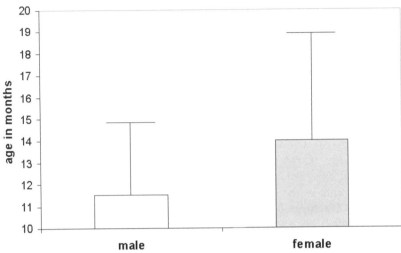

Fig. 4: Minimum age in months between both sexes (mean ± SD); females: n = 36, males: n = 55, p = 0.003.

Whereas the main proportion of males disappeared in the course of their second year (94.12 %), more females reached their third year of life (23.08 %) compared to males (5.88 %): Only two males lived until the third year while six females could be observed until the third year and one of them was still alive in autumn. The majority of male individuals disappeared until the end of spring (55.88 %). The maximum life span represents 2.42 years (29 months) in females and 2.17 years (26 months) in males.

4 Discussion

Our study population showed relatively high density levels between 2001 and 2005. RUZIC (1977) categorised different population density levels (from "very low" to "very high") on the basis of data from former Yugoslavia. Her classification supported the assumption that the population density in our site can be assigned to a "high" population. The peak population density in 2001 might be related to several building projects in the surroundings forcing the animals to emigrate into the study area. In addition, Common Hamsters are known to show occasionally population explosions within short time periods under favourable food conditions (GRULICH 1978, 1986). This might have been the case in our site, too. The very high density declined until 2003, probably due to lower carrying capacity of the habitat, but remained at a relatively high level. In more natural habitats, a maximum population density of 2.85 individuals/ha (WEIDLING & STUBBE 1997) and 3.6 individuals/ha (WEINHOLD 1998) were found. SELUGA (1996) found not more than 1 individual/ha on average. There were only small differences in population development between the sexes, indicating that the population decline might be rather due to limited resources than due to dispersal. To guarantee the continuity of a hamster population, WENDT (1989) stated that the spring population density should be between 0.5 and 2.0 individuals/ha. Accordingly, we considered our urban population as very stable. This is also supported by several reports and observations of residents observing hamsters at our site since the 1980s. Still, genetic exchange is necessary to maintain the population at a long-term basis. Isolated sites with no opportunities for exchange can probably not persist (GODMANN 1998).

Winter survival was determined by the number of individuals that had disappeared over the winter period. Analyses only included resident individuals shortly before hibernation. Frequent observations of the burrow and recapture attempts in spring were applied to measure over-winter survival accurately. The proportion of disappeared individuals during winter was quite high in both periods in our study. Since several authors found mortality rates between 40 % and 50 % (WEINHOLD 1998; KAYSER et al. 2003), it seems that over-winter mortality in this species is very high in general. WENDT (1991) even found a winter mortality of 61.5 %, which is in line with our results. However, compared to other species, Common Hamsters show very low survival rates over winter. For example, NEUHAUS and PELLETIER (2001) found an over-winter survival in adults of 90 % in Columbian ground squirrels (*Spermophilus columbianus*). In hedgehogs (*Erinaceus europaeus*), most deaths occurred during winter ranging from 26 % to 43 % in adults (KRISTIANSSON 1984). The winter mortality in adult hedgehogs was never higher than 40 % in southern Germany, (ESSER 1984). Over-winter mortality in Golden Marmots (*Marmota caudata aurea*) turned out to be lower than 10 % for adult individuals (BLUMSTEIN & ARNOLD 1998). The low winter survial rates in Common Hamsters found in our and other studies may probably be related to several aspects: Body mass shortly before hibernation was not found to influence over-winter survival in adult hamsters (FRANCESCHINI-ZINK & MILLESI 2008). SAINT GIRONS et al. (1968) stated a minimum necessary prehibernatory body mass of 150 g in juvenile hamsters. In juveniles having survived over winter, body mass shortly before hibernation was always above 160 g at our site. Still, several juvenile hamsters did not survive the winter. Moreover, according to some studies hamsters hibernated successfully even when the prehibernatory body mass was lower than 150 g (KRAMER 1956; SELUGA 1996). Hence, there is no evidence that physical condition before hibernation is an indicator of successful over-winter survival in this species. Common Hamsters do not only rely on their body fat reserves during winter. They build up food caches from which they are able to feed from during winter. WENDT (1991) pointed out that high winter mortality may be related to disturbed food storing activity in autumn. It was neither possible to qualify nor to quantify the food caches in the burrow in our study population. But, it seems plausible that hamsters with larger food storages have better chances to survive the winter. High winter mortality could also be related to the quality of food caches because in contrast to more natural habitats, hamsters were not able to cache durable food like cereals in Vienna. Furthermore, producing litters late in the season might shorten the time for food caching mainly in females. Therefore, over-winter survival might also be affected by reproductive effort and timing in the previous season. Further analyses will be necessary to support this assumption.

In hibernating mammals more juveniles than older individuals suffer death during winter (ESSER 1984; BLUMSTEIN & ARNOLD 1998; BERTOLINO & CURRADO 2001) due to lack of suitable hibernacula, food stores (WENDT 1991; SELUGA 1996), and less time for prehibernatory fattening. But this was not the case in our study. Juvenile hamsters were found to have similar over-winter mortality rates as older individuals. Reproductive activity may play an important role in this context because juveniles were found to participate in reproduction only in very exceptional cases in their first season (TSCHERNUTTER unpubl. data). It can be assumed that they need time for growth and preparation for hibernation and therefore delay in most cases reproduction to the second year. Besides, juvenile hamsters, born early in the year, might have better chances to survive over winter because they have more time to grow and prepare for hibernation (SELUGA 1996). Unfortunately, we do not have enough data to support this. In Vienna, Common Hamsters breed from May on, and females remain reproductively active until mid September (FRANCESCHINI-ZINK & MILLESI 2008), whereas males regress their testes beginning at the end of July (LEBL 2005). This long mating period in both sexes associated with an enormous temporal and physical investment might negatively affect over-winter mortality in adult compared to juvenile individuals. Reproductive effort might also explain why adult females suffered higher mortality over winter than adult males. Although hamsters are temporally limited, females manage to produce up to three litters per season. These physical demands might negatively influence female survival as found in several other species (SAETHER 1988; MARTIN 1995; NEUHAUS & PELLETIER 2001). However, up to now we cannot support a trade-off between female reproductive output and over-winter survival. But it seems that reproduction might be strongly involved in over-winter survival in this species.

Finally, we compared over-winter mortality between both winter periods and found small differences: In the winter 2003/04 more hamsters died than in the 2004/05 winter. An increasing rat population (*Rattus norvegicus*) in the study area may have had a negative effect. Rats are potential food and habitat competitors and seem to be able to displace hamsters (FRANCESCHINI & MILLESI unpublished) from their original habitat. During the winter 03/04 frequent feeding by residents led to an extremly high rat population density. Torpid hamsters might also function as a food resource for rats during the winter as observed in the Netherlands. There is some evidence for a relationship between low over-winter survival in hamsters and rat abundance, when these species inhabit the same area (MÜSKENS pers. comm.). This could explain higher survival rates in winter

2004/05 when we recorded a lower rat abundance compared to the winter 2003/04. In hibernating hedgehogs (*Erinaceus europaeus*) an interaction between over-winter mortality and the abundance of several rodent species has been observed (HOECK 1987).

Analyses on turnover rates give a reliable estimate about population fluctuations within a study area. In Vienna, high seasonal turnover rates (> 0.8) were recorded in male and female hamsters, indicating that almost every individual has been replaced in the course of a season. The results of our study are in line with findings from SAMOSH (1972) who observed that Common Hamster populations are completely replaced within two years. Similar results were found by WEIDLING & STUBBE (1997) (0.9) and WEINHOLD (1998) (0.87) when comparing spring and autumnal population composition. In our study, the seasonal turnover values for males were higher than those of females indicating sex differences in dispersal behaviour. Higher surface activity and larger home ranges are characteristic for male Common Hamsters (KAYSER 2001) and for males in several other species (e.g. GAULIN & FITZGERALD 1988, SHIER & RANDALL 2003). Following KUPFERNAGEL (2003), we set traps at a distance of 700 m. But none of the disappeared hamsters could be relocated. Up to now, we still have no information on the source of the immigrating animals. Especially in spring most male hamsters showed high fluctuation rates. This could be related to the search for oestrous females in other adjacent areas. But dispersing animals are also more exposed to risks than non-dispersers. Mainly in urban habitat, street crossings constitute a serious barrier leading to several victims. But this effect turned out to be rather density-dependent: In 2001, the year with the highest population density, several individuals suffered death caused by traffic. During the following years with lower densities almost no killed hamsters were found in the proximity of streets (FRANCESCHINI pers. obs.). Hence, dispersal may have been disproportionally stronger under high density conditions (KREBS 1992; SINCLAIR 1992) as intraspecific competition rises and resources are exhausted. Unfortunately, in most cases we could not determine the sex. Therefore, we cannot confirm higher losses due to traffic in males compared to females. Another reason for high male losses in spring might be due to the timing of the active season: Males appear earlier above ground than females (FRANCESCHINI & MILLESI 2005), which probably creates the disadvantage of low food availability, no predation protection due to low vegetation height and consequently high losses. We therefore assume that seasonal mortality in males might be basically affected by dispersal and predation.

Seasonal turnover rates were even higher compared to over-winter turnover rates. Our findings support results of a previous study (WEIDLING & STUBBE 1997) also showing by far higher seasonal turnover rates than over winter turnover rates. In both studies it could not be distinguished between mortality and dispersal during the active season. However, turnover rates during winter seem to be a good measure to estimate over-winter survival, as Common Hamsters stay resident shortly before hibernation. Therefore, dispersal effects are negligible when analysing over-winter survival. Nevertheless, several other factors may be involved in higher turnover rates during the season compared to the winter period: One reason for these seasonal differences might be that there are fewer factors that affect survival during hibernation than during the active season, when environmental influences (reproduction, predation) are plentiful (NEUHAUS & PELLETIER 2001). During the active season, hamsters are strongly exposed to predators. This effect might be enforced by low vegetation height, which is given within the whole study site. Birds of prey (*Falco tinnunculus*), Martens (*Martes foina*) (FRANCESCHINI & MILLESI unpublished), Foxes (*Vulpes vulpes*) and feral cats have been observed to be the most important predators. Foxes and martens are probably more likely to kill adult hamsters, whereas kestrels and feral cats focuse on juvenile hamsters. Injuries are estimated to be mainly caused by predators assuming that the influence of fights among conspecifics leading to death plays a subordinate role in this context. High predation pressure during summer could be a possible explanation for relatively low winter mortality (MILLESI et al. 1999): High predation pressure led to high losses in Ground Squirrels (*Spermophilus citellus)* in summer. This probably reduced competition for suitable hibernacula and food resources among the survivors.

Both high seasonal and high over-winter mortality rates in Vienna might be responsible for a relatively short life span in hamster. We found a maximum age of 2.42 years for females and 2.17 years for males. Recent field studies are in line with our results and suggest a much lower longevity than four years as stated by several other authors (SAINT-GIRONS et al. 1968; KLEVEZAL & KLEINENBERG 1969; VOHRALIK 1975): By means of capture-mark-recapture methods WEIDLING & STUBBE (1997) could follow individual hamsters at most over 11 months indicating that the majority of hamsters only live for one season. KAYSER (2003) observed a life expectancy persisting over a maximum of three active seasons. JONES (1982) found a maximum age of 2 years and 10 months in captivity. The sexual dimorphism in life span is a quite common phenomenon in several species (e.g. HOFFMANN et al. 2003; DAVIES et al. 2005) and also well known in humans (e.g. SMITH 1993; TERIOKHIN et al. 2004). In our study this sexual dimorphism was also detected. In Common Hamsters, as in other rodent species, larger home ranges in males compared to females (KUPFERNAGEL 2003) might be associated with a higher mortality risk (MICHENER 1989). Dispersing animals are exposed to a greater extent to risks leading to higher mortality than in non-dispersers (FERRERAS et al. 2004). These factors could explain the sex difference in maximum and minimum age. In contrast, the Syrian Hamster (*Mesocricetus auratus*) is remarkable among mammals because males tend to outlive females (KIRKMAN & YAU 2005).

Our results clearly show that this species is able to maintain high population densities over several years in an urban environment. Demographic parameters are similar to those in more natural habitats. High turnover rates mainly during the active season suggest that genetic exchange is sufficient. This is an important fact as the problem of habitat fragmentation may be high in urban areas. A wide spectrum of predators in the urban habitat indicates similar predation pressure as in natural habitats. In sum, high reproductive output and immigration seem to be responsible for the relatively stable density in the investigated Common Hamster population in Vienna. Nevertheless, urban populations still remain very sensitive to any kind of human influence that causes habitat loss.

5 Acknowledgements

Research was supported by the Austrian Science Fund (FWF Project 16001/B06) and the Austrian Academy of Science (Project 21930). We are grateful to E. SCHMELZER, C. ADLAßNIG, B. TAUSCHER, I. TSCHERNUTTER, K. LEBL, G. ROISER-BEZAN and C. SIUTZ for their help and assistance in field. We thank an anonymous reviewer for his/her helpful comments.

6 References

BACKBIER, L.A.M., GUBBELS, E.J., SELUGA, K., WEIDLING, A., WEINHOLD, U. & ZIMMERMANN, W. 1998: Der Feldhamster *Cricetus cricetus* (L., 1758), eine stark gefährdete Tierart. — Proceedings of the 4th Meeting of the International Hamsterworkgroup, Limburg.

BAUER, K. 1989: Rote Listen der gefährdeten Vögel und Säugetiere Österreichs und Verzeichnisse der in Österreich vorkommenden Arten. — Wien: Österrreichische Gesellschaft für Vogelkunde.

BERTOLINO, S. & CURRADO, I. 2001: Ecology of the Garden Dormouse (*Eliomys quercinus*) in the alpine habitat. — Trakya Univ. J. Scient. Res. B **2**: 75–78.

BLUMSTEIN, D. & ARNOLD, W. 1998: Ecology and Social Behavior of Golden Marmots (*Marmota caudata aurea*). – J. Mamm. **79**: 873–886.

DAVIES, S., ROSHANI, K., BHAIRAVI, B., PETHERWICK, A. & CHAPMAN, T. 2005: The effect of diet, sex and mating status on longevity in Mediterranean fruit flies (*Ceratitis capitata*), Diptera: Tephritidae. — Exp. Gerontol. **40**: 784–792.

ESSER, J. 1984: Untersuchung zur Frage der Bestandsgefährdung des Igels (*Erinaceus europaeus*) in Bayern. — Ber. ANL **8**: 22–62.

FERRERAS, P., DELIBES, M., PALOMARES, F., FEDRIANI, J. M., CALZADA, J. & REVILLA, E. 2004: Proximate and ultimate causes of dispersal in the Iberian lynx *Lynx pardinus*. — Behav. Ecol. **15**: 31–40.

FRANCESCHINI, C. & MILLESI, E. 2001: Der Feldhamster (*Cricetus cricetus*) in einer Wiener Wohnanlage. — Jb. Nass. Ver. Naturkde. **122**: 151–160.

FRANCESCHINI, C. & MILLESI, E. unpublished: Influences on population development in urban living European Hamsters (*Cricetus cricetus*). — Proceedings of the 11th Meeting of the International Hamsterworkgroup 2005, Budapest.

FRANCESCHINI, C. & MILLESI, E. 2005: Reproductive timing and success in European hamsters (*Cricetus cricetus*). — Proceedings of the 12th Meeting of the International Hamsterworkgroup, Strasbourg.

FRANCESCHINI-ZINK, C. & MILLESI, E. 2008: Reproductive performance in femal common hamsters. — Zoology **111**: 76-83.

GAULIN, S.J.C. & FITZGERALD, R.W. 1988: Home-Range Size as a Predictor of Mating Systems in Microtus. — J. Mamm. **69**: 311–319.

GODMANN, O. 1998: Zur Bestandssituation des Feldhamsters (*Cricetus cricetus* L.) im Rhein-Main-Gebiet. — Jb. Nass. Ver. Naturkde. **119**: 93–102.

GRULICH, I. 1978: Standorte des Hamsters (*Cricetus cricetus* L., Rodentia, Mamm.) in der Ostslowakei. — Acta Sc. Nat. Brno **12**: 1–42.

GRULICH, I. 1986: The reproduction of *Cricetus cricetus* (Rodentia) in Czechoslovakia. — Acta Sc. Nat. Brno **20**: 1–56.

HOECK, H.N. 1987: Hedgehog mortality during hibernation. — J. Zool. **213**: 755–757.

HOFFMANN, I.E., MILLESI, E., HUBER, S., EVERTS, L.G. & DITTAMI, J.P. 2003: Population dynamics of European ground squirrels (*Spermophilus citellus*) in a suburban area. — J. Mamm. **84 (2)**: 615–626.

JONES, M.L. 1982: Longevity of captive mammals. — Zool. Garten N.F. **52**: 113–128.

KAYSER, A. 2001: Aspekte der Raum- und Baunutzung beim Feldhamster. — Jb. Nass. Ver. Naturkde. **122**: 149–150.

KAYSER, A. 2003: Survival rates in the Common hamster. — Proceedings of the 10th Meeting of the International Hamsterworkgroup, Tongeren, Belgium: 105–108.

KAYSER, A., WEINHOLD, U. & STUBBE, M. 2003: Mortality factors of the common hamster *Cricetus cricetus* at two sites in Germany. — Acta Theriol. **48**: 47–57.

KIRKMAN, H. & YAU, P.K.S. 2005: Longevity of male and female, intact and gonadectomized, untrected and hormone-treated, neoplastic and non-neoplastic syrian hamsters. — Am. J. Anatomy **135**: 205–219.

KLEVEZAL, G.A. & KLEINENBERG, S.E. 1969: Age determination of mammals from annual layers in teeth and bones. —Israel Progr. Sci. Transl., Jerusalem.

KRAMER, F. 1956: Über die Winterbaue des Hamsters (*Cricetus cricetus* L.) auf zwei getrennten Luzerneschlägen. — Wiss. Z. Univ. Halle, Math.-Nat. V/4: 673–682.

KREBS, C.J. 1992: The role of dispersal in cyclic rodent populations. — In STENSETH, N.C. & LIDICKER, W.Z. (eds.): Animal dispersal. Small mammals as a model. pp. 160–175. — London Chapman and Hall.

KRISTIANSSON, H. 1984: Ecology of a hedgehog *Erinaceus europaeus* population in southern Sweden. — PhD Thesis University of Lund, Sweden.

KUPFERNAGEL, C. 2003: Range utilisation of translocated common hamsters *Cricetus cricetus* (Linnaeus, 1758) on a compensation area near Braunschweig. — Braunschweiger Naturkundl. Schriften **6**: 875–887.

LEBL, K. 2005: Physiologie und Verhalten bei männnlichen Feldhamstern (*Cricetus cricetus*) in Abhängigkeit von Alter und Kondition. — Master thesis, University of Vienna, Austria.

LENDERS, A. & PELZERS, E. 1985: Some data on the presence of the Common hamster *Cricetus cricetus* (L. 1758) in or near man-made objects in the Netherlands. — Lutra **28**: 2.95–96.

MARTIN, T.E. 1995: Avian life history evolution in relation to nest sites, nest predation, and food. — Ecol. Monogr. **65**: 101–127.

MICHENER, G.R. 1989: Sexual differences in inter-year survival and life span of Richardson's ground squirrels. — Can. J. Zool. **67**: 1827–1831.

MILLESI, E., HUBER, S., EVERTS, L.G. & DITTAMI, J.P. 1999: Sex and age differences in mass, morphology, and annual cycle in European ground squirrels, *Spermophilus citellus*. — J. Mamm. **80**: 218–231.

MÜHLENBERG, M. 1993: Freilandökologie. — Heidelberg, Wiesbaden: Quelle und Meyer Verlag München, 3. Auflage.

NEUHAUS, P. & PELLETIER, N. 2001: Mortality in relation to season, age, sex, and reproduction in Columbian ground squirrels (*Spermophilus columbianus*). – Can. J. Zool. **79**: 465–470.

NEUMANN, K. & JANSMAN, H. 2004: Polymorphic microsatellites for the analysis of endangered common hamster populations (*Cricetus cricetus* L.). — Cons. Gen. **5**: 127–130.

RUZIC, A. 1977: Study of the population dynamics of common hamsters (*Cricetus cricetus* L.) in Vojvodina. — Beograd: Plant Protection **28**: 289–300.

SAETHER, B.-E. 1988: Pattern of covariation between life-history traits of European birds. — Nature **331**: 616–617.

SAINT GIRONS, M.CH., MOURIK, W.R. & VAN BREE, P.J.H. 1968: La croissance ponderale et le cycle annuel du hamster, *Cricetus cricetus canescens* Nehring, 1899, en captivité. — Extrait de Mammalia **32**: 577–602.

SAMOSH, V.M. 1972: Growth and development of *Cricetus cricetus* L. — Vest. Zool. **4**: 86-89.

SELUGA, K. 1996: Untersuchungen zu Bestandssituation und Ökologie des Feldhamsters, *Cricetus cricetus* L., in den östlichen Bundesländern Deutschlands. — M. Sc. Thesis, Martin-Luther-University Halle-Wittenberg.

SHIER, D.M. & RANDALL, J.A. 2003: Spacing as a predictor of social organization in kangaroo rats (*Dipodomys heermanni are*nae). — J. Mamm. **85**: 1002–1008.

SINCLAIR, A.R.E. 1992: Do large mammals disperse like small mammals? — In STENSETH, N.C & LIDICKER, W.Z. (eds.): Animal dispersal. Small mammals as a model. pp. 229–242. — London Chapman and Hall.

SMITH, D.W.E. 1993: Human longevity. — New York: Oxford University Press.

SPITZENBERGER, F. 1999: Verbreitung und Status des Hamsters (*Cricetus cricetus*) in Österreich. — In STUBBE, M. and A. (eds.): Ökologie und Schutz des Feldhamsters, pp. 111–118. — Halle/Saale, Wissenschaftliche Beiträge der Martin-Luther-Universität Halle-Wittenberg.

TERIOKHIN, A.T., BUDILOVA, E.V., THOMAS, F. & GUEGAN, J.F. 2004: Worldwide variation in life-span sexual dimorphism and sex specific environmental mortality rates. — Hum. Biol. **76**: 623–641.

VOHRALIK, V. 1975: Postnatal development of the Common hamster *Cricetus cricetus* (L.) in captivity. — Rozpravy CSAV **85**: 1–49.

WEIDLING, A. 1997: Zur Raumnutzung beim Feldhamster im Nordharzvorland. — Säugetierkd. Inf. **21**: 265–273.

WEIDLING, A. & STUBBE, M. 1997: Fang-Wiederfang-Studie am Feldhamster *Cricetus cricetus* L. — Säugetierkd. Inf. **21**: 301–310.

WEINHOLD, U. 1998: Zur Verbreitung und Biologie des Feldhamsters (*Cricetus cricetus* L. 1758) in Baden-Württemberg – unter besonderer Berücksichtigung der räumlichen Organisation auf intensiv genutzten landwirtschaftlichen Flächen im Raum Mannheim-Heidelberg. — Phd Thesis, University of Heidelberg.

WENDT, W. 1989: Feldhamster, *Cricetus cricetus* (L.). — In STUBBE, H. (ed.): Buch der Hege I – Haarwild, pp. 667–684. — Harry Deutsch Verlag, Frankfurt/Main.

WENDT, W. 1991: Der Winterschlaf des Feldhamsters *Cricetus cricetus* (L., 1758) – Energetische Grundlagen und Auswirkungen auf die Populationsdynamik. — Wiss. Beitr. Univ. Halle: Populationsökologie von Kleinsäugerarten 1991: 67–78.

WOLLNIK, F. & SCHMIDT, B. 1995: Seasonal and daily rhythms of body temperature in the European hamster (*Cricetus cricetus*) under semi-natural conditions. — J. Comp. Physiol. B **165**: 171–182.

ADDRESSES OF THE AUTHORS:

CLAUDIA FRANCESCHINI-ZINK
Department of Behavioural Biology
University of Vienna
Althanstraße 14
A-1090 Vienna, Austria
claudia.franceschini@univie.ac.at

EVA MILLESI
Department of Behavioural Biology
University of Vienna
Althanstraße 14
A-1090 Vienna, Austria
eva.millesi@univie.ac.at

Role of the Common Hamster (*Cricetus cricetus*) in the diet of natural predators in Hungary

ZOLTÁN BIHARI, MÁRTON HORVÁTH, JÓZSEF LANSZKI & MIKLÓS HELTAI

Abstract: Common Hamsters (*Cricetus cricetus*) living in agricultural lands are a very important prey species of many predatory birds and mammals in Hungary. In some areas its high densities, make it the main prey of certain strictly protected predators. Hamsters prefer open agricultural areas, and are frequently preyed upon by predators foraging in these habitat types. Hamsters were found in the pellets of 14 bird species and in the stomachs of 7 carnivore species in Hungary. In some regions, during gradation, hamster remains made up 70–80 % of the pellets in the Eagle Owl (*Bubo bubo*) and Imperial Eagle (*Aquila heliaca*), making it the main prey of these strictly protected animals. The importance of the Common Hamster as a prey species varies, it changes between years of gradation and also between the different gradations. Our study suggests that the relatively rare predators have no serious impact on hamster populations, but that hamsters can play a very important role in the diet of these strictly protected predators. The protection of these predators is therefore highly dependent on the conservation of Common Hamsters.

1 Introduction

Predation can be a very important mortality factor in Common Hamster populations (KAYSER et al. 2003). The main causes of mortality of hamsters in Hungary are (in order of estimated importance): (1) infections at the peak of gradation (can kill most of the overflowed population); (2) extreme weather conditions; (3) pest control by poisons; (4) intensive agricultural methods (drastically changing the habitats and food supplies); (5) increasing traffic; and finally (6) predation, which can important especially in small and peripheral populations, causing fragmentation (BIHARI & ARANY 2001).

In Germany, KAYSER et al. (2003) found that mainly raptors (*Milvus migrans, Milvus milvus, Buteo buteo* and *Aquila pomarina*) are preying on hamsters, but these rodents can play an important role in the diet of certain carnivores as well (*Vulpes vulpes, Mustela erminea, Meles meles, Canis lupus* f. familiaris). Juvenile hamsters can be caught by further avian predators (*Falco tinnunculus, Ardea cinerea, Corvus corone corone, Corvus frugilegus*).

The hamster can reach high local densities, making it a key prey item for many predators foraging in agricultural areas. The present study discusses the available data on predation on the Common Hamster in Hungary and examines the hamster's role in the diet of predators by analysing the temporal relative frequency of it's occurrence in their diets.

2 Methods

All available data on the diet of potential predators of Common Hamsters in Hungary were collected from the literature. These studies were based on pellet- and stomach analyses and field observations.

The feeding habits of two endangered and one common predator species were examined. In addition, stomach analyses were conducted on samples of carnivores, such as the Steppe Polecat (*Mustela eversmanni*, n = 95), European Polecat (*Mustela putorius*, n = 44), Weasel (*Mustela nivalis*, n = 155), Badger (*Meles meles*, n = 35), Domestic Cat (*Felis catus*, n = 264) and Red Fox (*Vulpes vulpes*), provided by the National Carnivore Monitoring Programme (BEGALA et al. 2000). Prey were analysed by microscope based on feather, bone, dental and hair characteristics using standard procedures (for details see: BÍRÓ et al. 2005). The diet of the Imperial Eagle (*Aquila heliaca*) was studied based on food remains and pellet analyses. 1,334 food remains and pellets of the Imperial eagle were collected in 11 different regions between 1997 and 2004.

3 Results

In Hungary the presence of Common Hamsters was verified in the diet of 14 bird and 7 mammal species, although several other predators probably prey regularly on hamsters as well.

3.1 Predatory birds

The Buzzard (*Buteo buteo*) is the most common bird of prey in agricultural fields. Nevertheless, only few data are available on its predation of hamsters. BALOGH & VARGA (1983) found 3 sousliks, 2 hamsters, 42 other rodents, 2 pigeons and 10 frogs and reptiles in the nests of buzzards. ANDRÉSI & SÓDOR (1987) found hamster remains in buzzard pellets. KALOTÁS & PINTÉR (1992) and VARGA & RÉKÁSI (1993) reported dead hamsters in the nests of buzzards. Hamsters composed 5.1 % of the buzzards diet were composed by hamsters (n = 78, PAPP 2001).

The Short-toed Eagle (*Circaetus gallicus*) is a rare bird in Hungary. On one occasion, a dead hamster was found in its nest (VARGA & RÉKÁSI 1993). The same authors found a dead hamster in a Goshawk (*Accipiter gentilis*) nest. BITTERA (1914) investigated the stomach of many rare birds and found that the

Montagu's Harrier (*Circus pygargus*) also preyed on hamster. The stomachs of two hunted Golden Eagles (*Aquila chrysaetos*) contained hamster remains (CHERNEL 1909). STERBETZ (1975) found dead hamsters under the nest of a White-tailed Eagle (*Haliaeetus albicilla*), and 11 dead specimens were found under the nests of Black Kite (*Milvus migrans*). Pellets of a Steppe Eagle (*Aquila nipalensis*) contained 8 hamsters (VÁNYI 1987). Hamsters are probably a very important food source for this particular bird because numerous remains were found under its roosting tree.

Imperial eagles regularly prey on hamsters, which represented 20–59 % of the prey items in the 11 regions studied (Fig. 1). Between 1989 and 1990 there was a gradation of hamsters, followed by a population collapse. From 1994 the population increased again, and a break out was detected afterwards in 1998–2000. The pellet analyses demonstrated that hamster predation was low between the two gradation peaks, whereas it was high during the gradation peak (Fig. 2). In those regions where eagles preyed more on hares (*Lepus europaeus*), the rate of hamster predation was lower. During hamster gradation the predation on the hare population decreased, which is advantageous from the game management perspective. Hamsters, as a temporal alternative prey, can be important for other large mammal and bird predators as well.

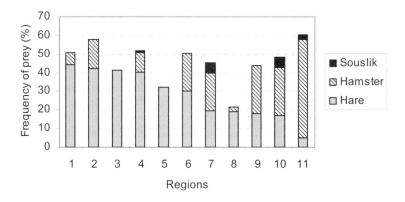

Fig. 1: Proportion of mammal prey of Imperial Eagles (*Aquila heliaca*) in 11 regions of Hungary (n = 1334) (1–Western Heves County, 2–Jászság, 3–Békés County, 4-Eastern Heves County, 5–Nagykunság, 6–Borsodi-síkság, 7–Bükk Mountains, 8–Mátra Mountains, 9–Börzsöny Mountains, 10–Eastern Zemplén Mountains, 11–Western Zemlén Mountains).

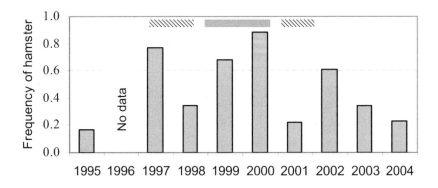

pre- and postgradation: ░░░░░░░ gradation peak: ▓▓▓▓▓▓

Fig. 2: Frequency of hamsters in the diet of Imperial Eagles (*Aquila heliaca*) in the Zempléni Mountains, Hungary.

3.2 Owls

The diet of the widely distributed Barn Owl (*Tyto alba*) was investigated intensively, but hamsters were not common prey items in Hungary. SCHMIDT & SÍPOS (1970–1971) found 14 hamsters in the pellets (0.2 % of the total number of prey). MOLNÁR (1983) found only one specimen in samples collected in western Hungary. ENDES & HARKA (1998) collected pellets in 50 villages, but reported hamsters in only 3 cases. BALOGH (1989) detected 3 specimens in pellets collected in eastern Hungary, which was 3.3 % of the total number of prey (n = 91). SCHMIDT (1969, 1971, 1976) found hamster remains in 26 owl pellets. Many unpublished data about hamster predation by barn owls are available, indicating that the hamster is a regular but not very common part of the barn owl's diet.

Only one case of hamster predation by a Long-eared Owl (*Asio otus*, SCHMIDT & SZABÓ 1981) has been reported.

The hamster is a very important prey of the Eagle Owl (*Bubo bubo*). The proportions of hamsters in pellets collected in 4 different regions of Hungary were 6, 20, 22.7 and 69.7 % (HARASZTHY 1984). Only the Rat (*Rattus norvegicus*) and the Hedgehog (*Erinaceus concolor*) are of equal importance for the owl. Another investigation (HARASZTHY et al. 1989) found proportions of 5.1, 7.6, 12, 24.1, 33.3, 34.2 and 60.8 % in the diet of different Eagle Owl territories.

3.3 Other birds

Non-predators occasionally also prey on hamsters. In Hungary, White Storks (*Ciconia ciconia*) often search for prey on agricultural fields after ploughing tractors, where they can easily catch injured, dead or young hamsters (NAGY 1964–1965). Rooks (*Corvus frugilegus*) also search for food on plough-lands, and RÉKÁSI (1973–1974) found hamsters in rooks' stomachs.

3.4 Carnivores

Red fox live in the same habitat as the hamster and are potentially a main predator. Hamsters remains have often been found around fox dens. The predation peak was in autumn, when hamsters are most abundant (Fig. 3).

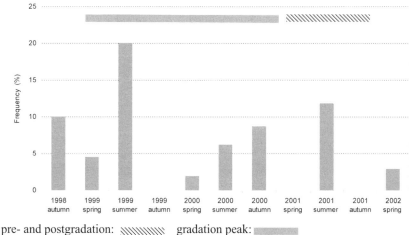

pre- and postgradation: ＼＼＼＼＼   gradation peak: ▬▬▬▬▬

Fig. 3: Proportion of hamster in the diet of Red Fox (*Vulpes vulpes*) (n = 247) in Hungary.

In Hungary, stray and feral cats and dogs often hunt on agricultural fields. Cats may prey on young hamsters, but only one observation is known (CSABA 1937). BÍRÓ et al. (2005) found 0.5 % hamsters in feral cat diets (n = 264), but no predation by wild cat (*Felis silvestris*) or hybrid cat. In one case the authors observed that a stray dog killed a hamster.

Hamster was identified in the stomach of Weasels (*Mustela nivalis*) only once (n = 155 stomachs). This small mustelid primarily feeds on voles, and can only predate young hamsters.

The Badger (*Meles meles*) probably eats dead hamsters, rather than directly hunting individuals, although it can dig up their nests. Only once (1 %) was hamster consumption proven by stomach analysis of badgers (n = 35).

The European Polecat (*Mustella putorius*) prefers rural habitats, so only three observations of hamster predations are available.

The Steppe Polecat (Mustella eversmanni) is a typical predator on agricultural areas of the Hungarian Great Plain, and the hamster is a main prey species. Hamster trappers often catch polecats with traps, positioned at the entrance of hamster burrows. This suggests that polecats often visit the burrows. Our investigation corroborated these observations: hamsters were the second most important prey species after the field vole (*Microtus arvalis*) (Fig. 4). Hamsters represented 19.6 % (n = 95) of the diet of steppe polecats.

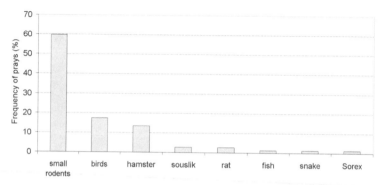

Fig. 4: Ranking of prey in the diet of the Steppe Polecat (*Mustella eversmanni*) in Hungary.

4 Discussion

The hamster is a very common rodent and important prey species in the Hungarian Great Plain. The distribution of Common Hamsters may have a serious impact on the distribution of three rare predators: Imperial Eagle, Steppe Polecat and Eagle Owl, whose ranges overlap with those of hamsters in Hungary. In the gradation years, hamsters can be their most important prey. Protecting rare predators requires the conservation of main prey items. Although carnivores and raptors can change their diet, they prefer hamsters when available over hares, a phenomenon with important consequences for game managers. Predation is a negligible factor for hamsters in the gradation years. Trappers, for example, catch about one million hamsters per year in Hungary, and pest control kills many millions during such years. Between gradations, however, when the density drops, predation could have a significant impact. Hamster populations can also be threatened by predators at the edge of the distribution area, where no gradations are noticeable. Here, the incidental hamster consumers can also threaten the population.

5 Acknowledgements

The authors are grateful to the Imperial Eagle Working Group of MME Birdlife Hungary for collecting and to Dr. BÉLA SOLTI (Mátra Museum) for identifying the prey items of Imperial Eagles.

6 References

ANDRÉSI, P. & SÓDOR, M. 1987: Sopron és környékének kisemlős faunája I. — Soproni Szemle **XLI/3**: 211-225.

BALOGH, L. & VARGA, ZS. 1983: Egerészölyv (*Buteo buteo*) és héja (*Accipiter gentilus*) táplálkozási adatok Sopron környékéről. — Madártani Tájékoztató, júl.-dec.: 104-105.

BALOGH, P. 1989: Adatok Derecske élővilágához. — Calandrella **II/2**: 89-93.

BEGALA, A., LANSZKI, J., HELTAI, M. & SZEMETHY, L. 2000: Adatok néhány fontosabb ragadozó táplálkozásáról. A vadgazdálkodás Időszerű Tudományos Kérdései **1**: 28-37.

BIHARI, Z. & ARANY, I. 2001: Metapopulation structure of common hamster (*Cricetus cricetus*) in agricultural landscape. — Jahrbücher des Nassauischen Vereins für Naturkunde **122**: 217-221.

BÍRÓ, ZS., LANSZKI, J., SZEMETHY, L., HELTAI, M. & RANDI, E. 2005: Feeding habits of feral domestic cats (*Felis catus*), wild cats (*Felis silvestris*) and their hybrids: trophic niche overlap among cat groups in Hungary. — J. Zool. **266**: 187-196.

BITTERA, GY. 1914: Nappali ragadozó madaraink gyomortartalom-vizsgálata. — Aquila **21/1-4**: 230-238.

CHERNEL, I. 1909: Adatok a húsevő madaraink táplálkozásának kérdéséhez. — Aquila **16**: 145-155.

CSABA, J. 1937: Nagycsákány emlősfaunája. — Vasi Szemle **IV/1-2**: 14-18.

ENDES, M. & HARKA, Á. 1998: Adatok a Tiszai alföld kisemlősfaunájához bagolyköpet-vizsgálatok alapján. — A puszta **1/15**: 159-167.

HARASZTHY, L. 1984: Adatok az uhu (*Bubo bubo*) magyarországi táplálkozásviszon-yainak ismeretéhez. — Puszta **2/11**: 53-59.

HARASZTHY, L., MÁRKUS, F. & PETROVICS, Z. 1989: Újabb adatok az uhu (*Bubo bubo*) magyarországi táplálkozásáról. — Madártani tájékoztató, jan.-jún.: 6-9.

KALOTÁS, ZS. & PINTÉR, A. 1992: Öt fióka egerészölyv (*Buteo buteo*) fészekben. — Madártani tájékoztató, jan.-jún. 24.

KAYSER, A., WEINHOLD, U. & STUBBE, M. 2003: Mortality factors of the common hamster Cricetus cricetus at two sites in Germany. — Acta Theriol. **48/1**: 47-57.

MOLNÁR, I. 1983: Bagolytáplálkozási adatok a Dunántúlról. — Madártani Tájékoztató, júl.-dec.: 106-110.

NAGY, I. 1964-1965: Adatok a gólya táplálkozásához (gólya hörcsög-pusztítása). — Aquila LXXI-LXXII: 231.

PAPP, G. 2001: Az egerészölyv (*Buteo buteo*) költésbiológiája - síkvidéki és hegyvidéki fészkelők összehasonlítása. — Debreceni Egyetem. 28 pp.

RÉKÁSI, J. 1973-1974: Adatok a vetési varjú (*Corvus frugilegus*) táplálékához Bácsalmás környéki mezőgazdasági területeken. — Aquila **80-81**: 291-292.

SCHMIDT, E. 1969: Adatok egyes kisemlősfajok elterjedéséhez Magyarországon, bagolyköpetvizsgálatok alapján. — Vertebrata Hungarica **XI/1-2**: 137-153.

SCHMIDT, E. 1971: Kisemlős-faunisztikai adatok Debrecen környékéről és az ország egyéb pontjáról bagolytáplálék-vizsgálatok alapján. — Múzeumi Kurír **6**: 21-26.

SCHMIDT, E. 1976: Kleinsäugerfaunistische Daten aus Eulengewöllen in Ungarn. — Aquila **82**: 119-144.

SCHMIDT, E. & SÍPOS, GY. 1970-1971: Kleinsäugerfaunistische Angaben aus dem Hernádbecken auf Grund der Gewölluntersuchungen der Schleiereulen (*Tyto alba* /Scop./). — Tiscia **6**: 101-108.

SCHMIDT, E. & SZABÓ, L. 1981: Data to the small mammal fauna of the Hortobágy based on owl pellet examinations. — The fauna of the Hortobágy National Park: 409-411.

STERBETZ, I. 1975: Adatok a Mártélyi Tájvédelmi Körzet emlős- és halfaunájáról. — Állattani Közlemények **LXII/1-4**: 107-114.

VÁNYI, R. 1987: Pusztai sas (Aquila nipalensis) a Nagy-Sárréten. — Madártani Tájékoztató, ápr.-szept.: 54-56.

VARGA, ZS. & RÉKÁSI, J. 1993: Adatok az Észak-borsodi Karszton fészkelő ragadozómadarak táplálkozásához és állományváltozásaihoz az 1986-1991 közötti időszakból. — Aquila **100**: 123-136.

ADDRESSES OF THE AUTHORS:

ZOLTÁN BIHARI
University of Debrecen
4032 Debrecen, Böszörményi út 138
Hungary
bihari@agr.unideb.hu

MÁRTON HORVÁTH
MME BirdLife Hungary
1121 Budapest, Költő u. 21
Hungary

JÓZSEF LANSZKI
University of Kaposvár
7401 Kaposvár. P.O. Box 16
Hungary

MIKLÓS HELTAI
St Stephen University
2103 Gödöllő, Páter K. u. 1.
Hungary

Peak numbers of *Cricetus cricetus* (L.): do they appear simultaneously?

GÁBOR NECHAY

Abstract: Population numbers of the Common Hamster increase conspicuously in certain years. Whether these hamster-outbreaks are cyclic, as in several other animal species, remains unknown. This hamster mainly inhabits cultivated land, i.e. habitats under considerable human influence. With the expansion of agricultural cultivation, hamsters increasingly damaged crops periodically during the 19th and 20th centuries; since the 1980s, however, the species has become endangered throughout the western part of its range. Detection of population cycles is difficult due to various human activities including hamster control measures and the lack of long-term series of exact large-scale data on population numbers. A review of historical and recent information, and available data on peak numbers, reveal that *Cricetus* outbreaks apparently have an 8–11 year periodicity. The cyclic fluctuation in number occurs simultaneously throughout or in wide territories of the species' range, similarly to Lemmings (*Dicrostonyx groenlandicus, Lemmus lemmus, L. sibiricus*), Mouse (*Mus domesticus*), Snowshoe Hare (*Lepus americanus*) or Voles (*Microtus agrestis, Microtus arvalis*).

1 Introduction

Population dynamics and especially the cyclic fluctuation of numbers of several animal species are one of the most challenging issues in ecology (e.g. ELTON 1942; LACK 1954; KREBS & MYERS 1974; KORPIMÄKI et al. 2004). The Common Hamster is a well-known animal that today exclusively inhabits arable land. The amplitude of its population changes is remarkable, considering that it is one of the few rodents of the original forest-steppe and steppe zone that survived the transformation of its natural habitats into agricultural fields. The hamster's population and distribution area even increased in croplands. Nevertheless, very few researchers took an interest in *Cricetus* population fluctuations. Possible explanations include:

- the long periodicity of its fluctuation in number,
- lack of adequate long-term series of population data in certain parts of the species' range,
- the belief or preconception that such fluctuations can only be investigated in species and population living in natural habitats,
- in parts of Europe where hamster research is currently flourishing, the species became rare or endangered in the last decades of the 20th century,

• there is an incomprehensible disinterest to investigate hamster biology in the central and eastern part of its range, where vital populations still live, like in Hungary.

Cricetus population fluctuations, however, show signs of cyclic nature throughout the wide range of the species, even those hamsters living in western Europe, where *Cricetus* became endangered over the last decades. When, for example, a small population increase is observed in Alsace (France), then the numbers slightly increase elsewhere too, e.g. in Bavaria (Germany); in the hamster regions of Hungary, where *Cricetus* is still considered a pest, outbreaks occur. Thus, hamster populations seem to change in a synchronous manner. NECHAY et al. (1977) attempted to visualize this spatial and temporal synchrony, listing peak years of certain countries. GRULICH (1980) reviewed several data and quantitatively evaluated the changes of hamster numbers. Long-term data on hamsters, however, differ considerably and there is insufficient quantitative information about changes in density, e.g. burrows/hectare. The use of the latter is inappropriate due to the major variation reflecting sudden changes of cropland hamster migration between various crops, not to mention the impacts of trapping and control measures in most of the fields in hamster peak years (outbreaks). Nevertheless, the present study reviews available information from the last centuries on hamster peak numbers to reveal signs of spatial and temporal synchronism.

2 Material and Methods

The nature, quality and quantity, as well as units of measurement of information are rather different, hindering quantitative evaluation of data series. The only viable opportunity for a long-term survey on *Cricetus* population changes is to list pieces of information about years with outstandingly high numbers of hamsters, i.e. information that relates to high and very low numbers of the species.

There are some records of mass multiplication (outbreaks) of hamsters in the course of the 17th, 18th and 19th centuries. As in other disasters like cholera, extreme droughts, floods, invasions of locusts or mice, variola and unusual weather conditions, someone usually noted abundant hamsters, occasionally together with the hamster numbers killed by people. In Hungary, RÉTHLY (1970, 1998) collected and published such recordings in two bulky volumes. STUBBE & STUBBE (1998) and STUBBE et al. (1998) reviewed and published similar information along with data on hamsters caught over certain periods in central Germany. However, these data usually refer to a specific locality and to smaller regions. For example, numerous hamsters appeared in Hungary and Transylvania (today Romanian territory) in 1720, mice (most probably Field Voles) were observed in "never before seen mass" in Hungary in 1732, and "susliks and steppe mice scamper like chickens" around Debrecen (eastern Hungary) in 1790. In eastern and western Hungary, "the hamster-army gathered the corn on the cob even before ripening. By the end of

August, one third of the maize was gone..." in 1914. Even in the early 20ᵗʰ century, most of the available information was along similar lines. Some data, however, is available on hamsters caught at various localities of certain countries (Belgium, The Netherlands, France, Germany), much less data from Hungary and the neighboring countries, and hardly any information about population changes from the wide eastern part of the range (Ukraine, Russia, Kazakhstan).

Thus, available data and various information on peak numbers of *Cricetus* have been examined in the present study from 'traditional' hamster regions of Europe and its eastern distribution area (Germany, Austria, Hungary, Slovakia, Romania, Ukraine and Russia) as well as from the edges of its range like Limburg (Belgium, The Netherlands), Alsace (France), northern parts of Germany, Kazakhstan and Siberia. Years with peak *Cricetus* numbers have been listed in a matrix according to countries or regions, and coincidences (coinciding years) have been selected.

The term 'peak number' in this report means (1) old information on hamster outbreaks, i.e. old memos like 'countless hamsters', 'hamsters harvest grains', or 'lots of hamsters occurred', (2) the first highest number of hamsters caught or their abundance recorded in the field in a sequence of years where available. Field records are usually based on the number of burrows per one hectare. (3) Data on extremely high damage caused or area occupied by hamsters.

3 Results

Few data are available from the 17ᵗʰ, 18ᵗʰ and 19ᵗʰ centuries, except from the late 1800s. These refer mainly to the late summer and autumn period of one peculiar year, occasionally giving the number of hamsters killed or caught only in a smaller region, e.g. around a village. Sometimes information is also given about a decrease or disappearance of hamsters in the following spring. It may be concluded that *Cricetus* outbreaks passed rather quickly. No periodicity is evident, but without doubt not all outbreaks have been recorded. However, those that were noticed must have represented real outbreaks.

There are exceptions. STUBBE et al. (1998) present, for example, the report of the town Gotha on payments from 1817 to 1830 for those who caught and handed in hamsters. 111,817 hamsters were caught alone in 1817, then 88,044 from August 1818 to April 1825 (12,577 a year on average), 19,795 (1 April 1825–12 September 1826) 21,843 (12 September 1825–30 September 1828, i.e. 7,281 a year) and 14,519 (10 September 1828–29 September 1830, i.e. 7,259 a year). Thus, there was certainly a hamster outbreak in 1817, followed by years in which about 10 times fewer hamsters occurred. From 1825–1826 there was a slight increase (peak year again), which, in my opinion, might actually have been higher if no permanent control by trapping occurred. Similarly, the probably first order to control hamsters – and oldest set of data from 1591 to 1596 (STUBBE &

STUBBE 1998) – show that, following a decree by the Mühlhausen local government, a few thousand hamsters were caught year by year, but 41,224 of them in 1596. The latter was certainly an outbreak. Moreover, no information is available on what happened before 1591, i.e. the events that actually prompted the municipal council to issue an ordinance.

Since and during the second half of the 19th century, a general expansion of the *Cricetus* population can be observed. This includes an outbreak in Alsace (1884), a first appearance in Mecklenburg (the 1860s) and a first outbreak in Rheinland-Pfalz (several thousand hamsters caught in 1894). It also includes a first outbreak in Limburg (Belgium, The Netherlands) in 1879–1880, and peaks from Alsace to Hungary (1898–1900) followed by organized control of hamsters (catching of hamsters from year to year and even control with CS2 and other methods) everywhere, similar to later peaks over the 20th century.

These control measures probably significantly influenced the 'natural' population development and prolonged the peak years, resulting in local oscillations of hamster numbers. This is reflected by relatively abundant data on captures in the western part of the range since the first outbreaks in the 16th and 19th centuries (e.g. data of DUPONT 1932; KOVÁCS & SZABÓ 1971; GRULICH 1980; KALOTÁS 1988, pers. com.; BAUMGART 1996; KRÜGER & KRÜGER 1998; STUBBE & STUBBE 1998; STUBBE et al. 1998; THIELE 1998).

Despite all the problems in evaluating the information, certain periods of Cricetus outbreaks are evident (Table 1).

Table 1: Periods of *Cricetus* outbreaks

1879-1881	L(1, 2) SA-Th-S(4)
1888-1890	L(1, 2) Als(5) SA-Th-S(3, 4)
1897-1900-1901	L(2) Als(5) Rh-Pf(6) SA-Th-S(3, 4) M(7), H(8) RO(8) which was extended also to 1902-1910 in Rh-Pf and M
1903-1905-1907	L(2) Rh-Pf(6) SA-Th-S(3) M(7)
1910-1912	L(4) Rh-Pf(6) SA-Th-S(3) A(9)
1913-1915	L(3) H(8) RO(8)
1920-1924	SA-Th-S(3) H(8) RO(8) UA(10) RUS(8)
1929-1931	everywhere
1951-1953	SA-Th-S(3) A(9) H(8) RUS(11)
1957-1959	Als(5) Rh-Pf(6) SA-Th-S(3) A(9) H(8) SK(12) RO(8) UA(10) KZ(13) which continued also in 1960 in H, SK, RO,UA
1966-1968	L(2) Rh-Pf(6) SA-Th-S(3) A(9) H(8,14) SK(12)
1971-1974	A(9) H(15) SK(12, 16) RO(8) RUS(13)
1982-1984	Als(4, 5) Rh-Pf(6) SA-Th-S(4) A(9) H(15, 17)

Remarks

A (Austria), Als (Alsace, France), H (Hungary), KZ (Kazakhstan), L (Limburg, (Belgium, Netherlands), M (Mecklenburg), Rh-Pf (Rheinland-Pfalz), SA-Th-S (Sachsen-Anhalt, Thüringien and Sachsen), SK (Slovakia), RUS (Russia), RO (Romania) UA (Ukraine)

Numbers in ( ) refer to: (1) DUPONT 1932, (2) PELZERS et al. 1984, (3) STUBBE & STUBBE 1998, (4) STUBBE et al. 1998, (5) BAUMGART 1996, (6) THIELE 1998, (7) KRÜGER & KRÜGER 1998, (8) NECHAY et al. 1977, (9) SPITZENBERGER 1998, (10) GORBAN et al. 1998 (11) NERONOV & TUPIKOVA 1967, (12) GRULICH 1980, (13) BERDYUGIN & BOLSHAKOV 1998, (14) KOVÁCS & SZABÓ 1971, (15) NECHAY 1998, (16) TOTH 1974, (17) KALOTÁS 1988 (pers. comm.).

Recently, a renewed increase took place in Hungary in 1997, with a peak in 1998–2000 accompanied by intensive control. To my knowledge, the situation improved somewhat again in Alsace, Germany, Austria and Hungary in 2005.

Table 2: Some characteristics of Vole, Lemming, Snowshoe Hare population cycles and of mice and hamster outbreaks (adapted from KORPIMÄKI et al. 2004)

	Microtus agrestis	*Lemmus lemmus, L. sibiricus, Dycrostonix groenlandicus*	*Lepus americanus*	*Mus domesticus (Australia)*	*Cricetus cricetus*
Periodicity (years)	3-5	3-5	9-11	4-8	(5) 8-11
Amplitude (fold)	50-500	100-1000	20-50	100-2000	≥ 30-800 and even ≥ 2450
extent of spatial synchronism (km)	70-500	200-1000	500-1500	50-1500	300-4000
max. rate of population increase	7-8 a year	not known	3-4 a year	1.16/month	(4 a year)

Population numbers of *Cricetus* fluctuate similarly to certain rodents and other species even though the hamsters mainly live in non-natural habitats, specifically in arable land. On the contrary, these cyclic fluctuations in number are especially remarkable in man-made habitats (cropland). Cycles are spatially and temporally synchronous. Following a summary (KORPIMÄKI et al. 2004) on key characteristics of lemmings, hares, mice and voles population cycles (Table 2), some features of Cricetus population changes can be outlined as follows. The amplitude of density changes has been reviewed for example by GRULICH 1980; STUBBE & STUBBE 1998; NECHAY 2000. Densities during outbreaks varied from

30 to 800 burrows/ha, in extreme cases even over 2000 per hectare (Slovakia in 1971, TOTH 1974; GRULICH 1980). The rate of yearly population growth during the reproduction period was, according to a field study on the increasing population, 264.7 % in 1983 and 409.1 % (1984), then 27.0 % (1985) and 63.6 % in 1986 (KALOTÁS 1988, pers. comm.). Fluctuations since the late 19th century, when uninterrupted data on high numbers of *Cricetus* are available, show an 8-11 year periodicity that is synchronous over a range of more than 1,000 km and even over 4,000 km in 1929–1931 and 1971–1974 (Table 2). A circa 5 year periodicity can also be observed in smaller regions.

4 Discussion

Listing and comparing information on peak years according to country proved to be more difficult than originally thought.

1. The simplest aspect is compiling memos on high hamster numbers. The logical assumption here is that note was taken of such events only when hamsters appeared in unusual numbers.

2. The optimal data are those relating to first mass-occurrence of hamsters in certain (new) areas. These indicate expansion of the distribution area, e.g. at the edges of the range like the Netherlands (Limburg) in 1879–1880 (PELZERS et al. 1984), Mecklenburg in 1898 (KRÜGER & KRÜGER 1998), or from the 1930s in Kazakhstan, where *Cricetus* was previously so rare "that the people inhabiting the region did not recognize the species" (BERDYUGIN & BOLSHAKOV 1998). Likewise, data and maps on areal fluctuations are good tools to point peak years out. Unfortunately, such maps are usually unavailable, but in Hungary spring and autumn counts of hamster burrows have been mapped twice a year since the 1970s (NECHAY 1998).

3. Protruding aggregate numbers of annual hamster catches in certain regions of the range can also be easily used, e.g. in Alsace (BAUMGART 1996), in Germany (STUBBE & STUBBE 1998) and in the former Soviet Union (NERONOV & TUPIKOVA 1967). Captures in certain areas are partly exact figures based on money paid for each hamster by the (local) government and on figures of trappers or fur collecting/processing enterprises. Other estimates are only approximate. In all cases, only those figures can be used that are the first outstanding ones after a sequence of years of low figures. In Slovakia, for example, GRULICH (1980) refers to data of a fur-collecting firm between 1950 and 1960. Accordingly, for the year 1950, 200,000 animals were recorded, and the decreasing numbers in subsequent years have been omitted. The outstanding figure for 1960 was 100,000: this value was considered, but because data are missing from 1958 and 1959, the 1960 outbreak potentially already developed in 1959.

4. Peak years during the second half of the 20th century have been identified mostly based on captures. When evaluating these figures, the following must be considered:

    a. In exactly described cases, at least a one-year deviation should be taken into account because the data show either the number of hamsters caught in a particular year or the number of hamster furs processed. The latter can represent hamster numbers valid for the preceding year.

    b. When *Cricetus* populations decline, professional trappers extend their trapping-area and/or they increase trap number (trap-nights/days). However, the number of hamsters caught is registered in the original region by the fur-collecting firm. (Hungarian trappers occasionally cross into neighboring countries to find untouched areas. I was also often asked for information about good hamster regions between 1970 and –1980, when I had up-to-date knowledge on hamster occurrences countrywide). Thus, the intensity of catching can differ significantly in the same or extended area of trapping in the year of a hamster outbreak and in subsequent years. That also helps explain why memos and diaries on the activity of trappers yield smaller variances in hamster numbers at a certain place than the real situation (for such a diary and changes in trapping intensity see: STUBBE & STUBBE (1998).

5. Certain peak years in certain countries are also determined based on hamster damage assessment. Such estimates reflect the value and quantity of crops destroyed or the area of cropland damaged by hamsters, along with the area of hamster control (TOTH 1974; NECHAY et al. 1977; GRULICH 1980; STUBBE et al. 1998). Interpreting such information, especially the area of damaged cropland, requires considering that the information may actually pertain to peak hamster numbers in autumn of the previous year when they damaged winter crops. The damage, however, was accounted for the next year, which could already correspond to the decrease phase of the population cycle.

6. The 8–11 year cyclicity of *Cricetus* population fluctuations has been concluded based on the survey of available information as shown in Table 1. Although the coincidences range between 3 and 5 years, the cycles are conspicuous. Divergences and, in certain parts of the range, (local) prolongations of peak numbers are evident. The potential role of human interventions here cannot be calculated. In Russia, *Cricetus cricetus* is believed to have a 9 year or decennial population cycle (BERDYUGIN & BOLSHAKOV (1998). However, long-term and large-scale population surveys on *Cricetus* have not been carried out, even though the species is an excellent model for research on population cycles.

7. The obvious question is what could trigger the synchronism of *Cricetus* population cycles over a wide range? It is beyond the scope of this paper to

discuss the factors and processes that apparently drive these cycles. Further review and field research is highly recommended, especially in the eastern part of the range, where the Common Hamster can still be investigated in near-natural habitats. The spatio-temporal synchrony of its population changes certainly deserves more attention, and this work should also consider the impacts of human management measures and of numerous predators on hamster populations.

5 Summary

Population changes of the Common Hamster *Cricetus cricetus* are comparable to those of certain species with well-known population cycles, such as lemmings and voles. The amplitude of densities is wide, and changes in abundance are temporally synchronous over a wide range, with an 8–11 year periodicity.

6 References

BAUMGART, G. 1996: Le Hamster d'Europe (*Cricetus cricetus* L. 1758) en Alsace. — Rapport realisé pour l'Office National de la Chasse, Décembre 1996.

BERDYUGIN, K.I. & BOLSHAKOV, V.N. 1998: The Common hamster (*Cricetus cricetus* L.) in the eastern part of the area. — In Stubbe, M. & Stubbe, A. (eds.): Ökologie und Schutz des Feldhamsters, pp. 43–79. — Halle/Saale, Martin Luther Universität, Germany.

DUPONT, C. 1932: La propagation du hamster en Belgique. — Bull. Musée royal d'Histoir naturelle de Belgique **8**: 1–43.

ELTON, C.S. 1942: Voles, Mice and Lemmings. – Oxford: Clarendon.

GORBAN, I., DYKIY, I. & SREBRODOLSKA, E. 1998: What has happened with *Cricetus cricetus* in Ukraine? — In STUBBE, M. & STUBBE, A. (eds.): Ökologie und Schutz des Feldhamsters, pp. 87–89. — Halle/Saale, Martin Luther Universität, Germany.

GRULICH, I. 1980: Populationsdichte des Hamsters (*Cricetus cricetus*, Mamm.). — Acta Sc. Nat. Brno **14**: 1–44.

KREBS, C.J. & MYERS, J.H. 1974: Population cycles in small mammals. — Adv. Ecol. Res. **8**: 267–399.

KORPIMÄKI, E., BROWN, P.R., JACOB, J. & PECH, R.R. 2004: The puzzles of population cycles and outbreaks of small mammals solved? — BioScience **54**: 1071–1079.

KOVÁCS, V. & SZABÓ, L. 1971: The multiplication of the common hamster (*Cricetus cricetus* L.) in the County Hajdú-Bihar, the possibilities of its control and their evaluation. — Növényvédelem **7** (2): 77–80.

KRÜGER, J. & KRÜGER, R. 1998: Zur früheren Verbreitung des Feldhamsters in Mecklenburg/Nordostdeutschland. — In STUBBE, M. & STUBBE, A. (eds.): Ökologie und Schutz des Feldhamsters, pp. 245–249. — Halle/Saale, Martin Luther Universität, Germany.

LACK, D. 1954: The Natural Regulation of Animal Numbers. — Oxford Univ. Press, Oxford, UK.

NECHAY, G., HAMAR, M. & GRULICH, I. 1977: The Common Hamster (*Cricetus cricetus* [L.]); a Review. — EPPO Bulletin 7 (2): 255–276.

NECHAY, G. 1998: The state of the Common hamster (*Cricetus cricetus* L. 1758) in Hungary. — In STUBBE, M. & STUBBE, A. (eds.): Ökologie und Schutz des Feldhamsters, pp. 101–110. — Halle/Saale, Martin Luther Universität, Germany.

NECHAY, G. 2000: Status of hamsters *Cricetus cricetus*, *Cricetulus migratorius*, *Mesocricetus newtoni* and other hamster species in Europe. — Nature and environment **106**: 73 pp. — Strasbourg: Council of Europe.

NERONOV, V.M. & TUPIKOVA, N.V. 1967: How is indicated the population number of animals by the regional collection of furs (for example the common hamster). — Fauna and Ecology of the Rodents (Moscow University) **8**: 188–196.

PELZERS, E., COENDERS, F. & LENDERS, A. 1984: Enige achtergronden van de toename van Hamsters (*Cricetus cricetus* L.) in Zuid-Limburg 1879–1915. — Natuurhistorisch Maandblad **73** (11): 207–213.

RÉTHLY, A. 1970: Idõjárási események és elemi csapások Magyarországon 1701-1800-ig — Budapest: Akadémia

RÉTHLY, A. 1998: Idõjárási események és elemi csapások Magyarországon 1801-1900-ig. — Budapest: Országos Meteorológiai Szolgálat.

SPITZENBERGER, F. 1998: Verbreitung und Status des Hamsters (*Cricetus cricetus*) in Österreich. — In STUBBE, M. & STUBBE, A. (eds.): Ökologie und Schutz des Feldhamsters, pp. 111–118. — Halle/Saale, Martin Luther Universität, Germany.

STUBBE, M. & STUBBE, A. 1998: Der Feldhamster *Cricetus cricetus* (L.) als Beute von Mensch und Tier sowie seine Bedeutung für das Ökosystem. — In STUBBE, M. & STUBBE, A. (eds.): Ökologie und Schutz des Feldhamsters, pp. 289–325. — Halle/Saale, Martin Luther Universität, Germany.

STUBBE, M., STUBBE, A. & WEIDLING, A. 1998: Der Feldhamster im Spiegel von Presse und Ämtern, von Vernichtungs- und Naturschutzstrategien. — In STUBBE, M. & STUBBE, A. (eds.): Ökologie und Schutz des Feldhamsters, pp. 333–416. — Halle/Saale, Martin Luther Universität, Germany.

THIELE, R. 1998: Der Feldhamster in Rheinland-Pfalz. — In STUBBE, M. & STUBBE, A. (eds.): Ökologie und Schutz des Feldhamsters, pp. 197–208. Halle/Saale, Martin Luther Universität, Germany.

TOTH, S. 1974: The Calamity Occurrence of Hamster (*Cricetus cricetus* L.) in Eastern Slovakia in 19711972. — Sbornik UVTI, Ochrana rostlin **10** (7): 69–74.

ADDRESS OF THE AUTHOR:
GÁBOR NECHAY
Podmaniczky u. 27
H-2100 Gödöllő
Hungary
nechay@invitel.hu

Is the Common Hamster a good example for nature conservation efforts? — Critical reflections on the law on nature conservation in theory and practice

ULRICH WEINHOLD

Abstract: The conservation of the Common Hamster has proved to be complex and difficult not only in terms of field management but also in politics and jurisdiction. Most of the current conservation programs in Germany have only short-term perspectives concerning financial and political support. The laws on nature conservation in Germany only protect the so-called "nesting sites" but not explicitly the habitat of the Common Hamster, which remains highly endangered and on the brink of extinction. Therefore, habitat loss due to building and road construction is possible without coming into conflict with the law — if the destruction and/or disturbance of hamster burrows as such can be avoided. This interpretation is in contradiction to the known spatial biology of the species because recent field studies have shown that Common Hamsters use a large area both seasonally and annually.

1 Introduction

The Habitats Directive (92/43/EEC), adopted by the Council of European Communities on May 21, 1992, is a treaty that regulates the conservation and protection of biodiversity with the means of habitat and species conservation in order to create a coherent European ecological network known as Natura 2000. The Member States were given two years time to implement the Habitats Directive in their national laws on nature conservation.

Since the implementation of the Habitats Directive in May 1994, the Common Hamster has become one of the first "test species" to which the practical requirements and consequences of the Habitats Directive for Annex IV species have been applied. Species listed in Annex IV are of community interest and in need of strict protection. The Habitats Directive (article 12) prohibits:

a) all forms of deliberate capture or killing of specimens of these species in the wild;
b) deliberate disturbance of these species, particularly during the period of breeding, rearing, hibernation and migration;
c) deliberate destruction or taking of eggs from the wild;
d) deterioration or destruction of breeding sites or resting places.

Furthermore, Member States shall take the requisite measures to establish a system of strict protection for the animal species listed in Annex IV (a) in their natural range (article 12).

As a typical farmland species, the Common Hamster is often subject of impact assessments due to steadily increasing land use for building and road construction.

Hamsters are known to be vagile species, using several burrows per season (KARASEVA & SHILAJEVA 1965; GORECKI 1977; WEIDLING 1996; WEINHOLD 1998; KUPFERNAGEL 2003). GRULICH (1978) observed hamster trails of 300-700 m in length and found poisoned hamsters up to 500 m away from the location of the bait. WEIDLING (1997), WEINHOLD (1997, 1998) and KUPFERNAGEL (2003) used radio telemetry to study spatial behaviour. They observed average home range sizes of 1.6 ha for males and 0.4 ha for females; the animals travelled routes of up to 605 m, along mostly linear structures, within 60 minutes.

One should expect that, since 1994, the conservation of this species has reached an advanced level based on up-to-date biological data, and that the Common Hamster functions as an example for other Annex IV species. The following article assesses the current legal status of the species in Germany as well as standard procedures based on this status; it then compares these with new field data on spatial ecology.

2 Material and Methods

2.1 Laws studied

Bundesnaturschutzgesetz, BGBl 2002, 1193 (BNatSchG, German Federal Law on Nature Conservation).

BfN (German Federal Agency for Nature Conservation): Konkretisierung der Ruhe- und Fortpflanzungsstätten von Anhang IV-Arten, 2004 (Specification of resting and breeding sites for Annex IV-species of the Habitats Directive).

Habitats Directive—Council Directive 92/43/EEC of 21 May 1992 on the conservation of natural habitats and of wild fauna and flora.

2.2 Field studies

Hamsters were live trapped at two neighbouring study sites near the city of Mannheim in the Rhine-Neckar region in Baden-Württemberg, Germany, in the years 2003 and 2004. The study sites are separated from each other by the motorway A 656, connecting the cities of Mannheim and Heidelberg. Each study site is completely surrounded by motorways, railways and suburban settlements; their sizes are 87 ha and 111 ha. The average spring burrow density was 0.5 burrows/ha in 2003 and 0.2 burrows/ha in 2004. The average summer burrow density was 0.6 burrows/ha in 2003 and 0.3 burrows/ha in 2004.

In total 40 live traps (mesh wire, 35 x 10 x 10 cm) were positioned in front of the burrows and checked at dawn and dusk. Hamsters trapped were first anaesthetised with ether, then weighed, sexed, body measured and ear tattooed. The trapping season began in May and ended in October of each year. Trapping sessions of one week were carried out in monthly intervals.

Seasonal recapture distances were measured within one year between each recapture event of each individual. Annual recapture distances of each individual were measured between the last capture in 2003 and the first in 2004.

2.3 GIS analysis

The GIS software Idrisi (Clark University, USA) was used to conduct buffer analyses on mean recapture distances. Idrisi is a raster-based GIS, which works with so-called image files. Vector data has to be converted to raster data prior to spatial analysis. The coordinates of the hamster burrows were taken in the field with a portable GPS (Garmin Etrex) and then downloaded onto the computer and imported into Idrisi. After conversion into raster data, Idrisi then creates the set buffer width around each of the hamster burrows, yielding a cloud-shaped buffer image. The buffer method has already been used by BERBERICH (1988) to calculate home range sizes of red fox (*Vulpes vulpes*).

2.4 Legal status and standard procedures

The German law on nature conservation states (BNatSchG 2002):

- To refrain from avoidable intrusion on and/or disturbance of nature and landscape (§ 19 BNatSchG)

- To compensate unavoidable intrusion on and/or disturbance of nature and landscape (§ 19 BNatSchG)

Paragraph 19 is applicable to every type of landscape. In addition, paragraph 30 defines strictly protected habitats, which exclude any intrusion/disturbance in general. But even a lifting of the restrictions of paragraph 30 can be obtained, if a building project is of high public interest or if there is no other feasible alternative (§ 34 BNatSchG).

The specific (municipal, regional or national) agency of nature conservation responsible for the building project in question first has to evaluate whether an intrusion on nature is avoidable or unavoidable. Sometimes this has to be negotiated on the basis of reports and statements handed over by the petitioner. Then an agreement has to be made on the scale of compensation measures for unavoidable intrusions. Compensation of landscape usually does not take into consideration species protection and does not refer to the species inventory of the landscape as such. The "ecological" value of the specific habitat type (e. g. meadow, marshland, heath, farmland etc.) lost to building has to be assessed and compensated by land of equal or higher value. These procedures are highly standardised and carried out by landscape ecologists.

If knowledge of protected species on or near the planned building site exists, paragraph 42 has to be consulted.

Paragraph 42 reads as follows:

- Not to kill, hurt or catch strictly protected species or disturb, destroy and take away their nests, hiding or breeding sites (§ 42 BNatSchG)

The prohibitions and restrictions of paragraph 42 (species protection) can be lifted on demand, if a building project is of high public interest or the restrictions would result in severe hardship for the petitioner (§ 62 BNatSchG).

A lifting of the restrictions of paragraph 42 first requires an impact assessment for the species concerned and an evaluation of possible compensation measures.

3 Field results

111 hamsters were trapped in 2003 and 35 in 2004. In 2003 the recapture rate was 35 %, in 2004 29 %. Twenty-two of the 35 individuals trapped in 2004 had been already marked in 2003.

The capture-mark-recapture studies yielded seasonal and annual recapture distances. The samples did not pass the normality test (K-S-Test, $p < 0.05$). The mean seasonal recapture distance was 100 m (min. 6 m, max. 473 m, $s \pm 120$, Fig. 1). The mean annual recapture distance was 366 m (min. 31 m, max. 871 m, $SD \pm 260$, Fig. 2).

The mean seasonal recapture distance for females was 71 m ($SD \pm 93$) and for males 141 m ($SD \pm 143$).

Females had a mean annual recapture distance of 387 m ($SD \pm 250$), males 346 m ($SD \pm 280$). The sex-specific differences in seasonal and annual recapture distances were not significant (Mann-Whitney U-test, $p \geq 0.05$).

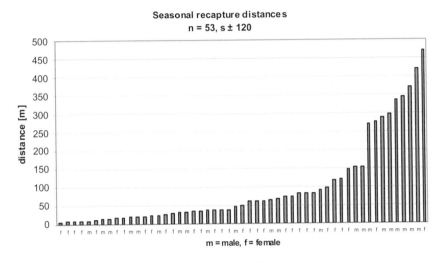

Fig. 1: Spectrum of seasonal recapture distances of Common Hamsters.

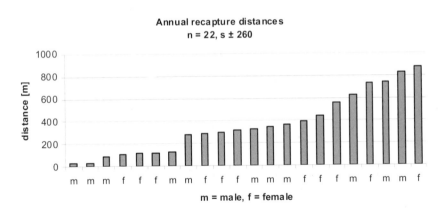

Fig. 2: Spectrum of annual recapture distances of Common Hamsters.

4 Conclusions

The German law on nature conservation does not refer to the species inventory of a habitat in terms of landscape protection and in terms of species protection; its efficiency is very limited and subject to interpretation because only nests, hiding and breeding sites are explicitly protected. Therefore, Common Hamster habitat can only be protected if an avoidable intrusion/disturbance is concluded by the relevant agency of nature conservation.

Farmland, as the presently typical Common Hamster habitat, has a very low ecological value in terms of landscape ecology and planning (BASTIAN & SCHREIBER 1999). It is not protected as such, comparatively cheap to acquire and easy to compensate for.

The interpretation guidelines of the Federal Agency of Nature Conservation (BfN 2004) for strictly protected species also define solely the burrow of the Common Hamster as the nest, hiding and breeding site. They, too, do not take the habitat into account. As a result, only the hamster burrows would have to be considered and referred to during planning and building procedures.

The observable clear but not significant difference between the mean seasonal recapture distance of females and males reflects the reproductive behaviour of this species. Female hamsters rear the young on their own and therefore are bound to a burrow for most of the reproductive period. Males, instead, do not invest in bringing up a litter but try to increase their reproductive success by mating with as many females as possible (EIBL-EIBESFELD 1953; WEINHOLD 1998). They therefore possess significantly bigger home ranges and much more actively roam their territory (WEIDLING 1997; WEINHOLD 1998; KUPFERNAGEL 2003).

The results of the field studies confirm the findings of WEIDLING (1997), WEINHOLD (1998) and KUPFERNAGEL (2003) and clearly show that hamsters migrate over considerable distances between burrows seasonally and annually. They use a large area over the year and from year to year. These migrations may be triggered by various environmental and seasonal factors as well as behavioural components like hibernation, reproduction, food supply and/or individual experiences not yet understood in detail. Patches of land may be occupied seasonally and/or annually, even if this is several hundreds of meters away from the currently populated area.

The frequently observed accumulation of hamster burrows (SELUGA 1996) in certain areas or fields therefore does not represent the complete habitat but is only a snapshot of the current seasonal situation.

The law on nature conservation, supported by the interpretation guidelines of the Federal Agency of Nature Conservation, in fact only requires testing the presence or absence of hamster burrows within the actual building site. Sometimes an additional buffer zone (area of disturbance) around the building site is monitored as well. If no hamster burrows are currently present, the laws do not have to be considered. Hamster habitat can therefore be destroyed without any species-specific compensation, even with a population nearby.

One key problem is the absence of an obligatory standard for the size of area to be monitored.

Furthermore, data on the overall distribution of the Common Hamster in a specific area are often lacking. Conclusions in terms of species protection according to § 42 are then often based on short-term and small-scale field results.

An approach to addressing the size of the area to be monitored for
a) an existing population would be a buffer analysis for each burrow with the mean seasonal and/or annual recapture distance as the buffer width (Fig. 3);
b) a building site within potential hamster habitat could be the same buffer analysis based on the building site boundaries (Fig. 4).

These approaches would be especially helpful if there is no other way to determine the boundaries of the overall habitat. In certain cases, the habitat boundaries are already given by human infrastructure (Fig. 5).

Another problem is the insecure financial situation of official conservation projects for the Common Hamster run by certain federal states of Germany like North-Rhine Westfalia, Baden-Württemberg and Bavaria.

Based on these findings, the conservation of the Common Hamster continues to suffer from a lack of professionalism because management and monitoring procedures are based on knowledge, which is insufficiently updated by field research. Moreover, the law on nature conservation has to be improved in terms of habitat protection for protected species in order to increase its efficiency. It would be desirable to officially support more field research in this matter and to improve the exchange of information between researchers and the authorities. Furthermore, the area of impact assessment should be based on the spatial requirements of the species and not on the artificial boundaries of the building sites.

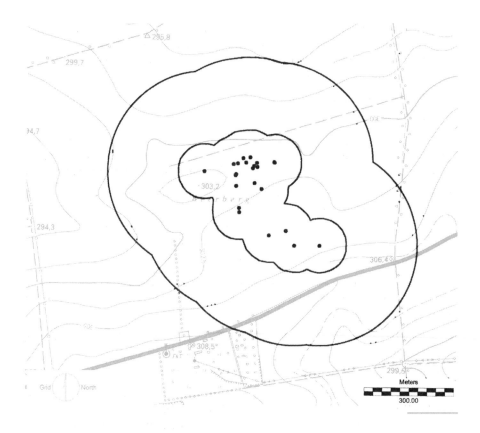

Fig. 3: Examples for buffer zones (black cloudy outlines) of a hamster colony (black dots) near the city of Erfurt (Thuringia, Germany). Inner circle = seasonal 100 m-buffer, outer circle = annual 366 m-buffer.

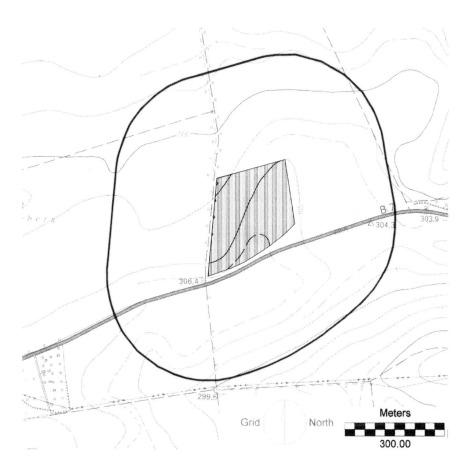

Fig. 4: Example for a 366 m-buffer zone (black outline) around the boundaries of a building site within potential hamster habitat, near the city of Erfurt (Thuringia, Germany, v-striped inner polygon).

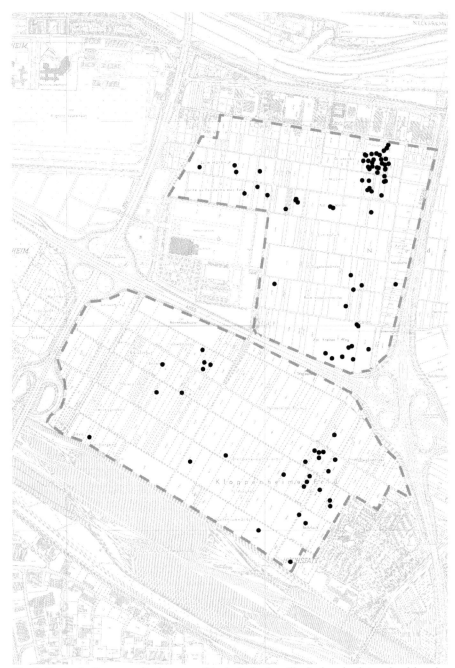

Fig. 5: Example for a monitoring area (black dashed outline) already determined by elements of human infrastructure near the city of Mannheim (Baden-Württemberg, Germany). Black dots = hamster burrows.

5 References

BASTIAN, O. & SCHREIBER, K.-F. 1999: Analyse und ökologische Bewertung der Landschaft, 2. Auflage. — Spektrum Akademischer Verlag.

BfN 2004: Konkretisierung der Ruhe- und Fortpflanzungsstätten von Anhang IV-Arten Entwurfsfassung zu ausgewählten Arten — Bundesamt für Naturschutz 2004.

BNatSchG 2002: Bundesnaturschutzgesetz (Gesetz über Naturschutz und Landschaftspflege) vom 25. März 2002.

BERBERICH, W. 1988: Untersuchungen zur Raumorganisation und zur Aktivitätsrhythmik des Rotfuchses (*Vulpes vulpes, L.*) im Hochgebirge. — Diss. Univ. Heidelberg.

EIBL-EIBESFELD, I. 1953: Zur Ethologie des Hamsters (*Cricetus cricetus L.*) — Z. Tierpsych. **10**: 204-254.

GORECKI, A. 1977: Ene

rgy flow through the common hamster population. — Acta Theriol. **22**: 25-66.

GRULICH, I. 1978: Standorte des Hamsters (*Cricetus cricetus L.*, Rodentia, Mamm.) in der Ostslowakei. — Acta Sc. Nat. Brno **12**: 1-42.

HABITATS DIRECTIVE — Council Directive 92/43/EEC of 21 May 1992 on the conservation of natural habitats and of wild fauna and flora.

KARASEVA, E.V. & SHILAYEVA, L.M. 1965: The structure of hamster burrows in relation to its age and the season — Bull. Moskauer Ges. der Naturforscher Abt. Biol. **70**: 30-39.

KUPFERNAGEL, C. 2003: Raumnutzung umgesetzter Feldhamster *Cricetus cricetus* (Linnaeus, 1758) auf einer Ausgleichsfläche bei Braunschweig. — Braunschweiger Naturkdl. Schriften **6**: 875-887.

SELUGA, K. 1996: Untersuchungen zu Bestandssituation und Ökologie des Feldhamsters, *Cricetus cricetus L.*, 1758, in den östlichen Bundesländern Deutschlands. — Diploma thesis, Univ. Halle-Wittenberg, Germany.

WEIDLING, A. 1996: Zur Ökologie des Feldhamsters *Cricetus cricetus L.*; 1758 im Nordharzvorland. — Diploma thesis, Univ. Halle-Wittenberg, Germany.

WEIDLING, A. 1997: Zur Raumnutzung beim Feldhamster im Nordharzvorland. — Säugetierkd. Inf. **4**: 265-273.

WEINHOLD, U. 1997: Der Feldhamster - ein schützenswerter Schädling? — Natur und Museum **127**: 445-453.

WEINHOLD, U. 1998: Zur Verbreitung und Ökologie des Feldhamsters (*Cricetus cricetus L.* 1758) in Baden-Württemberg, unter besonderer Berücksichtigung der räumlichen Organisation auf intensiv genutzten landwirtschaftlichen Flächen im Raum Mannheim-Heidelberg. — Dissertation, Univ. Heidelberg, Germany.

WEINHOLD, U.

ADDRESS OF THE AUTHOR:
ULRICH WEINHOLD
Institut für Faunistik
Rabelsacker 9
D-69253 Heiligkreuzsteinach, Germany
weinhold@institut-faunistik.net

Using track tubes to verify the syntopic occurrence of two ground-dwelling rodent species

ILSE E. HOFFMANN

Introduction

European Ground Squirrels (*Spermophilus citellus*) and Common Hamsters (*Cricetus cricetus*) have equal habitat requirements (SPITZENBERGER 2001), yet rarely are they observed to coexist in the same area. This fact is most likely due to resource partitioning by allocating surface activity to different temporal niches. While *S. citellus* is aboveground exclusively during open daylight after dawn and before dusk (EVERTS et al. 2004), *C. cricetus* prefers twilight and night hours (WENDT 1989). Burrow entrances of both species show very similar characteristics, such as the diameter (5–10 cm, cf. HUT & SCHARFF 1998), presence of soil mounds, and trails on the surface between the holes. Hence, such burrows disclose the presence of either one of these ground dwellers, but do not reveal the identity of their inhabitants. Counting only holes therefore constitutes a rather unreliable method to determine whether ground squirrels or hamsters, or both are present.

In the course of a previous mapping project (HOFFMANN 2002), it turned out that burrow entrances with the characteristics described above occur in a considerable proportion of habitats suitable both for hamsters and ground squirrels. A reference collection we had established from local species (Fig. 1) had revealed that *S. citellus* and *C. cricetus* tracks differ considerably in shape (Fig. 1, b, c). To examine the presence and particularly the abundance of each species, I therefore adapted track tubes (GLENNON et al. 2002) to gather footprints.

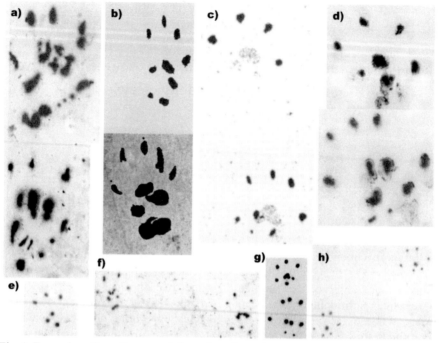

Fig. 1: Examples of small-mammal footprints from the reference collection. a) Eastern Hedgehog *Erinaceus concolor*, b) European Ground Squirrel *Spermophilus citellus*, c) Common Hamster *Cricetus cricetus*, d) Brown Rat *Rattus norvegicus*, e) Wood Mouse *Apodemus sylvaticus*, f) Bank Vole *Clethrionomys glareolus*, g) Eastern House Mouse *Mus musculus*, h) Common Vole *Microtus arvalis*. Each panel shows fore- (above) and hindpaw (below) except e) and h): only forepaws.

Track tubes were constructed from PVC-rain gutter tubes, cut into 33 cm long and approximately 10 cm wide and 8 cm high sections, and taped on aluminium plates. Clear adhesive film was placed on the aluminium plate with the sticky side up. Felt rectangles (9 x 7 cm) soaked with a mixture of carbon black and paraffin oil were fixed on both ends of the track plate and served as ink pads. Bait (nougat crème with oil, carrot pieces and breadcrumbs) was placed in the centre of each track plate. The tubes were positioned near burrow entrances and remained there for at least 30 hours. This covered the time span between dusk and the following dawn, and hence the period of aboveground activity in each species. Animals stepped onto the felt upon entering tubes and, by running through, transferred ink onto the surface of the adhesive film. I collected tracks by attaching the adhesive film to white paper, which enhanced visibility and served as a permanent record. Species were determined by comparing tracks with the reference collection.

Syntopy of ground squirrels and hamsters was first verified in an 8 ha area in the south of Vienna (48°09'N, 16°25'E, elev. 243 m). The site belongs to a former broadcasting station of the Foreign Service of Austria, and can be classified as semi-arid grassland dominated by *Bromus erectus* (HOFFMANN 2002). Except of mowing once or twice a year, the meadow is hardly exposed to any anthropogenic influence. The area is fenced, and surrounded by one-family houses in the North and East and vineyards in the South and West. We started to distribute track tubes close to burrow entrances on 6 May 2003, attempting to spread them evenly throughout the whole study area. The site was revisited on 8, 9, 11 and 13 May (intervals: 29–46 h) to check and re-position the tubes. All together, 49 entrances were checked this way.

Animals produced visible footprints in 51 % of the tracks tubes (Table 1). The majority was assigned to Ground Squirrels and Hamsters (72 % of track tubes with footprints). Of the remaining, three were left behind by eastern European Hedgehogs (*Erinaceus concolor*), three by Soricids or Murids, and one by a Snail. Two of the 18 tubes that showed footprints from *S. citellus* and / or *C. cricetus* had been entered by both species.

Table 1. Assignment of footprints collected in 25 of 49 track tubes to different sets of animals.

	Track tubes	
Type of tracks	#	%
S. citellus	12	48
C. cricetus	4	16
S. citellus and *C. cricetus*	2	8
Other	7	28

The coexistence of *S. citellus* and *C. cricetus* was again confirmed in 2004 during a capture-mark-recapture study on the same site. Although we focused then on European ground squirrels and adapted the timing accordingly (1000-1630 CET), Common Hamsters were caught in two cases late afternoon. This leads to the conclusion that track tubes detect the same species as those captured with conventional live traps. Track tubes may therefore provide an alternative sampling technique for documenting the presence of small mammal species with relatively low effort.

Acknowledgements

Thanks to CLAUDIA FRANCESCHINI-ZINK for collecting reference footprints and to ANNA STRAUSS for her assistance in the field.

References

EVERTS, L. G., STRIJKSTRA, A. M., HUT, R. A., HOFFMANN, I. E., & MILLESI, E. 2004: Seasonal variation in daily activity patterns of free-ranging European ground squirrels (*Spermophilus citellus*). — Chronobiol. Int. **21**: 57–71.

GLENNON, M. J., PORTER, W. F. & DEMERS, C. L. 2002: An alternative field technique for estimating diversity of small-mammal populations. — J. Mamm. **83**: 734–742.

HOFFMANN, I.E. 2002: Distribution of European ground squirrels in the southern districts of Vienna. — Report MA 22-3827/2002: Municipal Department for Environmental Protection, Vienna, 12. 10. 2007.

http://www.wien.gv.at/umweltschutz/pool/pdf/ziesel.pdf

HUT, R.A. & SCHARFF, A. 1998: Endoscopic observations on tunnel blocking behaviour in the European ground squirrel (*Spermophilus citellus*). Z. Säugetierk. **63**: 377-380.

SPITZENBERGER, F. (ed.) 2001: Die Säugetierfauna Österreichs.-Vienna: Grüne Reihe des BM für Land- und Forstwirtschaft, Umwelt und Wasserwirtschaft.

WENDT, W. 1989: Zum Aktivitätsverhalten des Feldhamsters, *Cricetus cricetus L.*, im Freigehege. — Säugetierkd. Inf. **3**: 3–12.

ADDRESS OF THE AUTHOR:

ILSE E. HOFFMANN

Department of Behavioural Biology

University of Vienna

Althanstraße 14

A-1090 Vienna

Austria

Tel.: ++43-1-4277-54469; Fax: ++43-1-4277-54506

ilse.hoffmann@univie.ac.at

Reproduction

How to increase the reproductive success
in Common Hamsters:
shift work in the breeding colony

STEFANIE MONECKE & FRANZISKA WOLLNIK

Abstract: Breeding and reintroduction programs of Common Hamsters are often impaired by the limited space in the breeding facility and by short time funding. The situation is aggravated by the short reproductive period of Common Hamsters, limiting breeding success to only one third of what could be achieved if the breeding period could be extended to a full year.

The reproductive cycle of Common Hamsters is controlled by an endogenous, circannual clock. It is synchronized to the natural year by an interaction of the animal's phases of sensitivity to short or long day information and seasonal changes in photoperiod. Thus, it is difficult to manipulate the duration and timing of the reproductive phase. Nonetheless, new insights into the neuroendocrine regulation of circannual rhythms and the ontogeny of Common Hamsters along with our long-term experience in breeding Common Hamsters has enabled us to develop a breeding regime that would allow breeding in a captive colony during most of the year. Our model also includes strategies to resynchronize the circannual rhythm of newborns to the natural year before release into the new habitat. We propose that our breeding regime would optimise the success of captive breeding colonies and, thus, of reintroduction programs.

1 Introduction

Common Hamsters (*Cricetus cricetus*) are critically endangered in Western Europe. In most Western European countries like France, Belgium, the Netherlands and Germany, the Common Hamster is strictly protected. Several measures are currently in place in order to increase the low number of individuals (NECHAY 2000). The last meeting of the international hamster workgroup (13th meeting, 14-17 October 2005) showed that neither this aim nor even a stabilisation of animal numbers could be achieved even though the small existing populations are additionally restocked by captive-reared animals. However, the breeding period of Common Hamsters is limited to only 4 months per year (NECHAY et al. 1977; KRSMANOVIC et al. 1984; SELUGA et al. 1996), lasting in captivity from early April to mid-August (REZNIK-SCHÜLLER et al. 1974; VOHRALÍK 1974; MASSON-PÉVET et al. 1994; MONECKE & WOLLNIK 2005). Therefore, the full capacity of breeding facilities can be utilized only during one third of the year. The model proposed here would help to use the breeding facilities more economically by providing a breeding regime framework that would allow breeding of Common Hamsters during most of the year. This would substantially increase the number of animals that can be reintroduced into the wild during one year.

Clearly, many of the litters would be born at the "wrong" time of year. Consequently, their reproductive and other seasonal cycles would be out of phase with the natural year. However, the correct synchronization of physiological functions to the natural seasonal cycle is essential for survival. The proposed breeding model, therefore carefully selects the timing and the natural conditions for reintroduction into the natural habitat, to allow a resynchronization of the pups' seasonal cycles.

1.1 The seasonal timing of reproduction

In all seasonally breeding species, including the Common Hamster, the reproductive cycle is timed such that the offspring is born in the most favourable seasons of the year, namely spring and summer. In Common Hamsters, the onset and offset of the breeding season are not passive reactions to external stimuli like seasonal changes in the climate, vegetation or day length. Instead, the reproductive cycle is driven by an endogenous circannual clock (MASSON-PÉVET et al. 1994) and persists even when the animals have no information about the time of year. In other words, the circannual clock induces both the development and regression of the gonads without external stimuli. Since the endogenous period length usually deviates slightly from 365 days, however, seasonal changes in photoperiod are necessary to synchronize the circannual clock to the external cycle of the natural year (GWINNER 1986; GOLDMAN 2001; ZUCKER 2001). In Common Hamsters these photoperiodic changes, so-called zeitgebers, are only effective when they coincide with one of two annual phases of sensitivity (SABOUREAU et al. 1999; MONECKE & WOLLNIK 2004). Around the summer solstice, between mid-May and mid-July, the animals are sensitive to a shortening of the photoperiod (SABOUREAU et al. 1999). During this time a decrease of day length below 15.5 to 15 h (at a latitude of 48°35' N) (CANGUILHEM et al. 1988) resets the clock and induces gonadal regression. From mid-November until early spring an increase in photoperiod above a critical value of more than 13 h of day length (at a latitude of 48°46' N) resets the clock, too, and induces gonadal development (MONECKE & WOLLNIK 2004).

1.2 The ontogeny of reproduction

To survive in a seasonally changing environment, adult Common hamsters undergo several physiological changes, avoiding the harsh winter conditions by hibernation and reproducing in spring and summer (MONECKE 2001). As the females have several litters in one year, the offspring is born at different seasons, namely between early May (mid spring) and late August (mid summer)

(VOHRALÍK 1974). Each pup must adapt its physiological state to the current season and synchronize properly to the cycle of the natural year. The offspring therefore shows a photoperiod-dependent timing of puberty (KIRN 2004). Litters, which are born early in the year and experience an increasing photoperiod, become reproductive in the same year. The first visible sign for the upcoming sexual maturation is the *descensus testes* in males and the opening of the vagina in females. In captive-bred Common Hamsters, born in increasing or long photoperiods, the descent of the testes can occur as early as the 21st day of life, and the opening of the vagina from the 34th day onwards (KIRN 2004). In contrast, litters born in a decreasing photoperiod become reproductive for the first time in the following spring, i.e. only after several months (TIEGS 2005). In contrast to Syrian Hamsters, which react to the ambient photoperiod only after puberty (GASTON & MENAKER 1967; DARROW et al. 1980; SISK & TUREK 1987), Common Hamster pups are well adapted to the ambient photoperiod from early life on. How do they synchronize to the natural year so early in life?

During gestation the photoperiodic information is transmitted from the mother to the embryos (HORTON 1984a, 1984b; STETSON et al. 1986; WEAVER & REPPERT 1986; HORTON & STETSON 1992). The current light dark cycle is perceived by the mother's retina. It has a direct (MOORE & LENN 1972) and an indirect (CARD & MOORE 1991) neuronal connection to the mother's circadian system, which is located in the suprachiasmatic nucleus (SCN) of the hypothalamus (KLEIN et al. 1991). The SCN measures the day length and mediates this information by a neuronal pathway to the pineal gland (MOORE 1996), where it is translated into the nocturnal peak of the hormone melatonin. The amplitude and duration of melatonin correspond to the night length (STEINLECHNER 1992; REITER 1993; SIMONNEAUX & RIBELAYGA 2003). This reliable intrinsic signal for the time of year is also conveyed to the embryos because melatonin passes the placenta (KLEIN 1972; REPPERT et al. 1979).

After birth the maternal influence on photoperiodic entrainment of the neonates is exiguous, as has been shown in other rodent species (ELLIOTT & GOLDMAN 1989; ROWE & KENNAWAY 2002). It is not known at which postnatal age rodent pups are able to perceive and to process photoperiodic information by themselves. Neuronal and molecular studies, however, show that the SCN, which measures day length, responds to light information before the eyes open (for review see DAVIS & REPPERT (2001)). The SCN of Common Hamsters starts to respond to light at an age of 9 to 14 days, based on light inducible *c-fos* expression (GERLING 2006). Thus, hamster pups probably perceive postnatal photoperiodic information at the latest at an age of 14 days, when the eyelids open (VOHRALÍK 1975). This information is then compared with the prenatal photoperiod mediated by the mother during pregnancy (KIRN 2004). Based on this comparison the offspring's puberty is stimulated in increasing or long photoperiods and is inhibited in decreasing photoperiods. Accordingly, Common Hamster pups become reproductive at an early age when they are born in spring and early summer, but postpone puberty to the next year when they are born clearly after the longest day, i.e., at a time when adults reach the end of the reproductive phase.

2 Results and discussion

2.1 Breeding model

As a circannual clock drives the reproductive rhythm of Common Hamsters, it is difficult to extend the breeding period in individual hamsters. However, the endogenous oscillation of the circannual clock can be reset during the animals' sensitive phases to photoperiodic information. The procedure is as follows: The breeding colony is divided into three groups. One group stays in natural photoperiodic conditions. The breeding phase will be – as usual – from early April to mid-August. For the two other groups the photoperiodic cycle is manipulated so that the reproductive phase is either advanced or delayed by several months. While the animals of the natural light group (NATURAL) can be kept outdoors, the advanced (ADVANCE) and delayed (DELAY) groups have to be placed in windowless rooms in which the artificial light onset and offset are controlled by a timer. The temperature can be either constant ($\leq 20°$ C) or adapted to the corresponding photoperiodic cycle. It is important to exclude the animals of the ADVANCE and DELAY group from any natural day light during maintenance in these conditions: even brief exposure to different light conditions can interfere with the entrainment to the new light conditions. Both shifted groups should consist of yearlings or – even better – juveniles born early in the current year.

2.1.1 ADVANCE group

The animals of the ADVANCE group are maintained in natural light conditions until June 1st, when they are transferred to an artificial constant winter photoperiod of LD09:15 (Fig. 1). At this date all animals are sensitive to a shortening of day length (SABOUREAU et al. 1999), which induces gonadal regression in yearlings (CANGUILHEM et al. 1988) and inhibits puberty in juveniles (KIRN 2004). In contrast to the natural shortening of day length, which affects Common Hamsters only in mid-July (SABOUREAU et al. 1999), the artificial short day signal of LD09:15 is effective in early June and advances the circannual cycle of the ADVANCE group by 1.5–2 months. Subsequently, the animals are maintained constantly in LD09:15 for several months. Though this is a winter photoperiod the ADVANCE group will become reproductive endogenously after a few months. Since their circannual clock is already advanced by 1.5–2 months and in many species, including the Common Hamster (MASSON-PÉVET et al. 1994), the period length of circannual rhythms is shorter than 365 d (GWINNER 1986), quite a few animals will become reproductive as early as December. Hence, the first litters should be ready for weaning 6 weeks later from mid-January on.

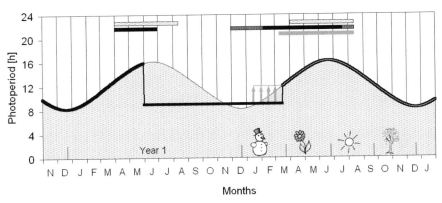

Fig. 1: Schematic representation of the photoperiodic conditions for the ADVANCE group (black line). These animals will become reproductive at the very beginning of their second year. The dotted area indicates seasonal changes in the natural photoperiod (here: for Stuttgart (latitude 48°46' N)), while the grey line indicates the photoperiodic regime for the offspring of the ADVANCE group after weaning. The bars at the top indicate the expected corresponding reproductive phases for the NATURAL group (dotted), the ADVANCE group (black) and its offspring (grey). The grey arrows indicate the time of transfer for the offspring.

At this date, the natural photoperiod might have already exceeded LD09:15, depending on the latitude of the breeding facilities, and the pups can be reintroduced into the natural habitat. If the natural photoperiod is shorter than LD09:15, or the winter is very cold, the pups can stay in the breeding facility. In any case, the pups of the ADVANCE group must not experience any decrease in day length because this would postpone puberty by several months. If the pups stay in the breeding facility, they should be maintained separately from the adults in a photoperiod of LD12:12 to subject them already to a clearly increased photoperiod (Fig. 1). However, LD12:12 is insufficient to stimulate puberty in juvenile hamsters (KIRN 2004), so that they need a further increase in day length to become reproductive. Thus, on March 21st, all weaned litters should be transferred either to a room with a window, to outdoor enclosures or to the field. At this day the natural photoperiod is LD12:12 worldwide and the subsequent increasing photoperiod will soon stimulate puberty. The parental generation should also be transferred to a natural photoperiod (room with a window) on March 21st. The new litters are born in a natural stimulatory photoperiod of >LD12:12 and will become reproductive early in life (KIRN 2004). These litters can be released into the field immediately after weaning.

In the ADVANCE group, all pups experience a clear increase of photoperiod rather early in the year at the time of reintroduction and are, therefore in phase with the natural reproductive cycle. The parental generation can also be reintroduced as soon as the breeding group in natural conditions becomes reproductive, i.e. in early April, because it is necessary to replace this group in the following year by newly shifted animals. The reintroduced parental generation can be mated in captivity in order to reintroduce pregnant females.

2.1.2 DELAY group

The DELAY group conditions are designed to breed non-reproductive juveniles, which can be reintroduced late in the natural year. The simplest strategy is to prevent the breeding animals for several months from experiencing the natural shortening of the photoperiod after the summer solstice (Fig. 2a). For this group the photoperiod of the longest day, June 21st, is extended to late October. Thus, the reproductive phase starts as usual in early April, but the end of the reproductive phase should be considerably delayed, even though probably not all animals will stay reproductive until late October due to the endogenous control of the reproductive cycle (MASSON-PÉVET et al. 1994). Our breeding facility, has had positive experience with this regime and some animals stayed reproductive even until the end of the year. Although this group can breed from early April on, mating should start only around the longest day, to avoid potential negative impact of too many pregnancies on the litter size at the end of the breeding period.

At weaning all pups are reintroduced to the wild or at least transferred to natural photoperiods in captivity. The associated decrease in day length will then inhibit gonadal development. From mid-September on, the adults should also be successively transferred to natural light conditions to alleviate resynchronization to the natural photoperiodic cycle. Mothers with pups are therefore transferred to natural light conditions at or soon after the birth of their last litter, and males when they are no longer needed for breeding in the current season.

This breeding model for the DELAY group is feasible in any breeding facility because it does not need complicated timer programs. It can be repeated every year. Problems may arise when the circannual clock induces gonadal regression endogenously in too many animals before intended by the breeding program. Moreover, if the animals stay reproductive, the unnaturally long duration of this physiological state might impair litter size. Such problems require modification of the program: a more complex lighting regime is needed, although it is sufficient to adjust the light timer once a week to the photoperiodic changes.

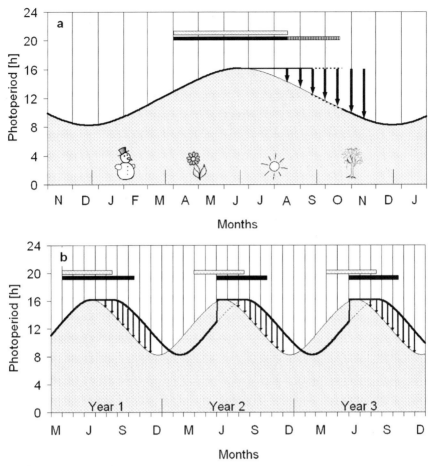

Fig. 2: Schematic representation of the photoperiodic conditions for the DELAY group
(black line) in which the reproductive period is either extended (a) or shifted (b)
by two months. The dotted area indicates seasonal changes in the natural photoperiod (here: for Stuttgart (latitude 48°46' N)). The bars at the top indicate the
expected corresponding reproductive phases for the NATURAL group (dotted)
and the DELAY group (black). The black arrows indicate the time of transfer for
the offspring.

In this modified model, the DELAY group is shielded from the natural decrease in day length for two months until August 21st. Only then, are the animals
released into a lighting regime mimicking the natural changes in photoperiod
with a 2-month delay (Fig. 2b). After synchronization to these conditions, the
animals' circannual clocks will be delayed. Thus, in the next year, they will become reproductive two months later so that breeding can start only in early June.
To prevent the pups, which would be born before the longest day in the delayed
conditions, from experiencing a stimulatory effect of the increasing photoperiod

during lactation, the DELAY group has to be transferred to the photoperiod occurring at the longest day at the very beginning of their reproductive period, i.e. on June 1st. On August 21st, when this group would normally reach the longest day, the animals are again released into a lighting regime mimicking the natural decrease in photoperiod until June 1st of the following year, when the cycle starts again. After synchronization to these conditions, the reproductive phase has a normal duration and the circannual clock of breeding animals is in phase with the artificial photoperiodic cycle. The breeding stock of the DELAY group has to be restocked from its own offspring each year.

Because the adults are transferred at the very beginning of the breeding period to long photoperiods, the offspring will never experience an increase in photoperiod before weaning. In these conditions, the pups are weaned and transferred to natural decreasing photoperiods between late June and late November. Consequently, their gonadal development is inhibited for several months until the naturally increasing photoperiod stimulates gonadal development in the following year. Thus, these pups are in phase with the natural cycle.

2.2 Breeding plan

The breeding model extends the 4-month-long natural breeding period by another 6 months. This enables breeding for species conservation programs nearly year round, i.e. at least from December to October, in three "shifts" of breeding groups. The number of litters and the necessary number of breeding pairs for the proposed model depends on the available mating boxes and the rotation rate at which the breeding pairs are exchanged in the mating boxes. At the breeding facilities of Rotterdam Zoo (HOFMEIJER & HOOGEVEEN 2005) and Stuttgart University (unpublished observations) mating typically occurred already at the first day after release into the mating box or at the latest within the following 4-5 days, according to the duration of oestrus cycle (REZNIK-SCHÜLLER et al. 1974). The exchange of the breeding pairs in the mating box can thus occur every 5 days, although an exchange every 7 days may be more practicable. The probability of fertile matings can be increased by keeping the breeding pairs together for 2 oestrous cycles, i.e. for 10 days.

At a rotation interval of 10 days and an overall breeding period of 46 weeks, 32 litters per mating box can be reared during one year by using the suggested model with three "shifts" of breeding groups. Thirty-two litters correspond to 208 pups, assuming an average litter size of 6.5 animals (see below). This number can be boosted to 299 pups in 46 litters per mating box if a 7-day rotation interval is chosen. In the latter conditions, 7 females (and up to 7 males) per mating box and breeding group are needed for optimal success. A rotation interval of 10 days requires only 5 females (and up to 5 males) per mating box and breeding group, assuming that each female is occupied with gestation (up to 21 days) and lactation (ca. 21 days) for 6 weeks (42 days). Additionally, at least 1 week for recovery should be allowed before renewed mating. In any case, during the 4-month breeding period per group, each female is mated 2 to 3 times. Three litters per year and female are well within the natural range (GRULICH 1986; NECHAY 2000; LEIRS 2002). Recently, however an outdoor study reported significantly larger litter-sizes in a female's first litter of a year compared to later born litters (TAUSCHER & et al. unpubl. data). In contrast, the litter sizes of most captive females in our breeding colony increased in the second and third litter (Table 1). This difference may reflect seasonal changes in food availability or temperature to which only wild females are exposed.

Table 1: Litter sizes (number of pups) in relation to litter number of individual females. The table includes all females that gave birth several times during one breeding season in our breeding facility in Stuttgart.

Mother	litter-size [pups/litter]			increase from 1st to 2nd litter [%]	increase from 1st to 3rd litter [%]
	1st	2nd	3rd		
# 15.6	5	9	10	180	200
# 23.7	8	8		100	
# 36.3	7	8		114	
# 37.5	8	5		63	
# 40.5	4	6		150	
# 42.5	7	6		86	
# 45.1	4	6		150	
# 57.2	8	8	10	100	125
# 61.6	8	8	9	100	113
Mean (± SEM)	6.6 (0.6)	7.1 (0.5)	9.7 (0.3)	115.8 (12.3)	145.8 (27.3)

2.3 Age of reintroduction from the physiological point of view

This breeding model is only reasonable, if the captive-bred pups are reintroduced into the field at or soon after weaning. In current species conservation programs however, Common Hamsters are only reintroduced as yearlings (U. WEINHOLD, Germany; M. LA HAYE, The Netherlands, personal communication). Since the breeding period in these facilities starts only rather late in the year, i.e. in May (DE VRIES 2002), many pups do not reach sexual maturation in their first year. Therefore, the captive bred animals are prevented from the high mortality risk in nature to which especially the small juveniles are exposed (KAYSER & STUBBE 2002), until they are reproductive in spring, when they are yearlings. Note, however, that the survival of captive-bred Common Hamsters reintroduced in the field as yearlings is very low: only 1 % of the males and 5 % of the females survive for one entire year (MÜSKENS 2005).

Reintroduced animals are subjected to a dramatic environmental change when transferred from indoor cages to outside fields. They have to cope with fluctuating abiotic factors and must learn how to find food and how to recognize and avoid predators. The low survival rates of only 1 % in male or 5 % in female yearlings suggest that most of the reintroduced animals are not fit enough for this task. This survival rate is considerable smaller than that of wild pups despite being similarly inexperienced to the risks outside like wild pups when they are leaving the maternal burrow and despite being exposed to a lower predation pressure than wild pups because of their larger size. In the Srostinsk region of the Altai territory (Russia) 8.7 % of the male and 16.4 % of the female pups were recaptured in the following year (KARASEVA 1962). In the Hakel area (Saxony Anhalt, Germany) the minimal survival rate for one year was 12 % in male and 8 % in female pups (KAYSER & STUBBE 2002). GÓRECKI (1977) found 15.8 % of the pups alive after one year in the valley of the Vistula River (Poland). Apparently natural born pups learn considerably faster or more efficiently to cope with the risks above ground than reintroduced yearlings.

Young hamsters leave the burrow for the first time at an age of about three to five weeks (EIBL-EIBESFELDT 1953; KARASEVA 1962; SELUGA et al. 1996). This fits quite well with the date of weaning in captivity. At weaning, captive-reared animals and naturally reared animals should have a similar horizon because both have experienced a comparably poor environment - either a poorly equipped cage or a dark burrow. Thus, at that age both should have a similar chance to survive. Wild animals might have only an exiguous lead of experience because they occasionally follow their mother up to the burrow entrance at an age of 2-3 weeks (EIBL-EIBESFELDT 1953; SELUGA et al. 1996). If captive-reared hamsters are held in captivity after weaning for further months, their stimulus-poor cage-environment hinders essential learning. Reintroducing of captive-bred pups soon after weaning, as the suggested breeding model proposes might thus considerably increase survival rates versus reintroduced yearlings. Research on survival rate and reproductive success in relation to reintroduction age would be crucial to elucidate which reintroduction strategy is more efficient.

Literature data are contradictory (for review see NECHAY 2000) on whether Common Hamster pups can reproduce in the year of birth. This raises doubts about whether a reintroduction of hamster pups is reasonable. Although marked pups were never observed to reproduce before hibernation (SELUGA et al. 1996; SCHMELZER 2005), reproduction in the year of birth cannot be excluded. Since only few pups were recaptured in the observed areas shortly after leaving the mother's burrow and since juveniles often disperse before they settle in an own burrow (SELUGA et al. 1996), the pups might have migrated to another area, where they could have raised litters. Conversely, other pups probably immigrated into the observed areas from neighbouring fields. However, newly marked animals have always been considered to be yearlings or adults. Precise aging of wild-caught Common Hamsters is extremely difficult. Biometrical data allows at least rough estimates of age. However, Common Hamster pups which become sexually mature show a remarkable increase in body mass: males can already reach a mass of about 350-400 g and females of 200–250 g at an age of 3 months (LEMUTH 2001; KIRN 2004). These values are not only within the normal range for sexually mature juveniles (GRULICH 1986) but also for yearlings (VOHRALÍK 1975; MASSON-PÉVET et al. 1994; MONECKE & WOLLNIK 2005). Biometrical data such as body mass can therefore easily lead to an overestimation of age in field studies.

KIRN (2004) has shown morphologically that Common Hamsters can reach sexual maturity soon after weaning. Such young animals are indeed fertile: VOHRALÍK (1974) reported on a female caught with a litter of 9 pups in August and weighing only 160 g. This individual was probably born in the current season and may have been 2 to 3 months old. NECHAY (2000) found 6 pregnant females with the upper M3 not fully developed, indicating a similar age. GRULICH (1986) used the abrasion of molars to estimate age and reported that 20 % of the females, born in the year of capture, had *maculae cyanae* indicating a past pregnancy. Moreover, about 14 % of male Common Hamsters of the same age trapped in June and July had completely developed testes as well as sperm in the *caput epididymis*. In our breeding colony, a 3-months-old male weighting 410 g (born and maintained in natural light conditions) was used successfully for reproduction. The resulting litter had 8 pups. This rules out any doubts that male or female Common Hamster pups can reproduce in their first year of life. Consequently, captive-born hamsters can be reintroduced into the field in their first year of life.

2.4 Seasonal timing of reintroduction from the physiological point of view

Because Common Hamsters are exposed to high predation pressure, only animals, that are reproductive or becoming reproductive (early-born litters in the NATURAL group and litters of the ADVANCE group) should be reintroduced before or during the above-ground season. The probability is quite high that these animals survive long enough to raise one litter in nature, because they can participate in the natural reproduction season immediately after reintroduction.

Also non-reproductive animals (e.g. those reared according to the DELAY group or the late-born individuals of the NATURAL group) can be reintroduced successfully so that they survive until the next reproductive period. For several reasons they should be reintroduced between late summer and early winter rather than at the end of the hibernation period. Non-reproductive animals are physiologically ready for hibernation (MONECKE 2001) and they will rapidly vanish into the protecting hibernacula for the next months. If the animals are provided with an artificial burrow in an unharvested field with a generous food supply for the winter, the mortality until the next spring should be negligible. A recent field study showed that the winter mortality in Common Hamsters is only 10 % of the overall mortality during summer (KAYSER et al. 2003) and affects mainly old or sick animals or those with inadequate food supply. The probability should be high that the released pups of the DELAY group or the late-born pups of the NATURAL group will survive until they leave their hibernacula in spring with completely developed gonads. Consequently, the reproductive success of these animals should be similar to that of ADVANCE group pups. The exiguous loss caused by winter mortality can easily be balanced by utilizing the free space in the breeding facility for breeding according to the DELAY and ADVANCE group model, yielding many additional animals for reintroduction. Autumn releases are therefore advantageous for the economical management of the breeding facility and for the reintroduction program.

Autumn reintroduction is not in contradiction with JORDAN (2001), who proposed spring as the best season for reintroduction, because then the population density is at its low (GÓRECKI 1977). If the non-reproductive animals are released only when the original population has entered hibernation, their impact on the original population structure will be effective only in spring and should thus be minimal.

A further advantage of the reintroducing non-reproductive animals in autumn is that they can habituate to abiotic factors and to living in a burrow before they are exposed to the high predation risk above ground. Moreover, hibernation is accompanied by a remarkable loss of memory due to a decreased neuronal connectivity (MILLESI et al. 2001). To ensure survival, hibernation is probably succeeded by a period of an increased capability of learning. Since the timing of awakening from hibernation differs individually, this important phase of learning might be missed if the animals are still in captivity.

2.5 Breeding at the wrong time of the natural year

The proposed breeding model is based on long-term experience in our hamster breeding facility. It has, however, never been tested in this form for reintroduction programs. This raises potential doubts about the reproductive success of animals in shifted conditions versus in natural conditions, and whether the model is worth testing.

These doubts are unfounded. Common Hamsters have been reared in Stuttgart since 1998, initially only in natural light conditions during the natural reproductive phase and later increasingly under artificial conditions nearly year round.

The sizes and the sex ratios of all 78 litters (541 animals) born so far in Stuttgart are shown in Fig. 3. Although not the result of a systematic study, they show some interesting trends: mean litter size was 6.9 ± 0.3 animals with a nearly balanced mean sex ratio of 51.7 ± 2.2 % males per litter. Under natural conditions, the sex ratio showed seasonal variations. Whereas more females were born at the beginning and end of the reproductive period, i. e. in March and July, the sex ratio clearly shifted in favour of males in June during the longest days (Fig. 3b). This effect was reproduced in artificial photoperiods, because the proportion of males was highest in LD16:08 and decreased with decreasing photoperiods (Fig. 3d). Interestingly, litters born beyond the natural reproductive period were often larger than those born between late March and late July in natural light conditions reaching an average litter size of 10.2 ± 0.8 animals in December (Fig. 3a). However, there was no seasonal variation in litter size under natural light conditions (Fig. 3b). Though the conditions in our breeding colony have remained unchanged over the years (temperature $18 \pm 2°C$, humidity 55 ± 5 %, food (Altromin 1314, breeding diet for rats, Lage, Germany), and water *ad libitum*, during gestation and lactation additional protein supply (baby food of the trademarks Milupa, Hipp, Humana) litter size varied slightly between years (Fig. 3c). In general, litters born in artificial photoperiods were slightly larger (7.6 ± 0.5) than those born in natural photoperiods (6.5 ± 0.4) (Fig. 3d), but both values are within the natural range (VOHRALÍK 1974; GRULICH 1986; NECHAY 2000). One litter has been bred in LD09:15 according to the conditions of the ADVANCE group (Fig. 3d). This litter consisted of 9 pups. Although higher litter sizes in artificial conditions might be accidental, the example does show that breeding in a winter photoperiod is possible. The authors would therefore like to encourage conservationists to test this breeding and reintroduction model. This model can probably be improved during the test period but, as such, provides a solid basis for increasing the breeding success of captive-breeding colonies.

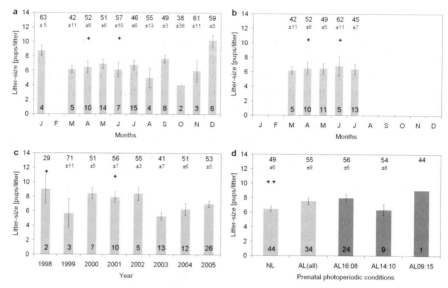

Fig. 3: Litter sizes and sex ratios in relation to the season of birth (a & b), to the year of birth (c) and to the prenatal photoperiodic conditions (d). Figures a, c and d include all litters, born in our breeding colony in Stuttgart between 1998 and November 2005 (n = 78) and reared either under natural light conditions (NL) or artificial light conditions (AL). Figure b shows only those litters born in natural light conditions (n = 44). For breeding in artificial conditions, three different photoperiods were used: LD16:08, LD14:10, LD09:15. Columns show litter size. Numbers at the bottom indicate the number of litters included in one column, numbers at the top the sex ratio [percentage of males per litter], and crosses two litters, that were excluded from the sex ratio analysis because pups died before sexing was possible. All data are presented as means ± SEM.

3 Acknowledgements

The authors wish to thank all co-workers of the department of animal physiology at the University of Stuttgart, who were involved in the breeding of Common Hamsters between 1998 and 2005: HARALD FEUCHTER, ANGELINE VOTTELER, NICOLE KIRN, HAGEN SCHMIDT, BIRGIT PELZ, ANDREA GERLING and KATHRIN CERNOCH. This study was supported by the German Research Foundation (DFG), WO 354/12-1.

4 References

CANGUILHEM, B., VAULTIER, J.-P., PÉVET, P., COUMAROS, G., MASSON-Pévet, M. & BENTZ, I. 1988: Photoperiodic regulation of body mass, food intake, hibernation, and reproduction in intact and castrated male Common hamsters, *Cricetus cricetus*. — J. Comp. Physiol. A **163**: 549-557.

CARD, J.P. & MOORE, R.Y. 1991: The organization of visual circuits influencing the circadian activity of the suprachiasmatic nucleus. — In KLEIN, D.C., MOORE, R.Y., REPPERT, S.M. (eds.): Suprachiasmatic nucleus - The mind's clock, pp. 51-76. — New York, Oxford: Oxford University Press.

DARROW, J.M., DAVIS, F.C., ELLIOTT, J.A., STETSON, M.H., TUREK, F.W. & MENAKER, M. 1980: Influence of photoperiod on reproductive development in the golden hamster. — Biol. Reprod. **22**: 443-450.

DAVIS, F.C. & REPPERT, S.M. 2001: Development of mammalian circadian rhythms. — In TAKAHASHI, J.S., TUREK, F.W., MOORE, R.Y. (eds.): Handbook of Behavioral Neurobiology (Vol. 12): Circadian Clocks, pp. 247-290. — New York: Kluwer Academic/Plenum Publishers.

DE VRIES, S. 2002: Breeding and reintroduction of the Common Hamster in the Netherlands. — In MERCELIS, S., KAYSER, A., VERBEYLEN, G. (eds.): Proceedings of the 10th Meeting of the International Hamster Workgroup, Tongeren (Belgium): pp. 42-43. — Natuurpunt Studie.

EIBL-EIBESFELDT, I. 1953: Zur Ethologie des Hamsters (*Cricetus cricetus*). — Z. Tierpsychol. **10**: 204-254.

ELLIOTT, J.A. & GOLDMAN, B.D. 1989: Reception of photoperiodic information by fetal siberian hamster: role of the mother's pineal gland. — J. Exp. Zool. **252**: 237-244.

GASTON, S. & MENAKER, M. 1967: Photoperiodic control of hamster testis. — Science **158**: 925-928.

GERLING, A. 2006: Entwicklung der c-Fos Expression im SCN des Europäischen Feldhamsters (*Cricetus cricetus*) [Development of *c-Fos* Expression in the SCN of European Hamsters (*Cricetus cricetus*)]. — Diploma Thesis, Biological Institute, Dept Animal Physiology, University of Stuttgart.

GOLDMAN, B.D. 2001: Mammalian photoperiodic system: formal properties and neuroendocrine mechanisms of photoperiodic time measurement. — J. Biol. Rhythms **16**: 283-301.

GÓRECKI, A. 1977: Consumption by and agricultural impact of the Common Hamster *Cricetus cricetus* (L.), on cultivated fields. — EPPO Bulletin 7: 423-429.

GRULICH, I. 1986: The reproduction of *Cricetus cricetus* (Rodentia) in Czechoslovakia. — Acta Sc. Nat. Brno **20**: 1-56.

GWINNER, E. 1986: Circannual Rhythms. — Berlin, Heidelberg, New York, London, Paris, Tokyo: Springer Verlag.

HOFMEIJER, J.K. & HOOGEVEEN, J. 2005: Breeding process of the European hamster, *Cricetus cricetus*, at Rotterdam Zoo, The Netherlands. — Talk at the 13th Meeting of the International Hamster Workgroup, Illmitz/Vienna.

HORTON, T.H. 1984a: Growth and maturation in *Microtus montanus*: effects of photoperiods before and after weaning. — Can. J. Zool. **62**: 1741-1746.

HORTON, T.H. 1984b: Growth and reproductive development of male *Microtus montanus* is affected by the prenatal photoperiod. — Biol. Reprod. **31**: 499-504.

HORTON, T.H. & STETSON, M.H. 1992: Maternal transfer of photoperiodic information in rodents. — Anim. Reprod. Sci. **30**: 29-44.

JORDAN, M. 2001: Reintroduction and restocking programmes for the Common Hamster (*Cricetus cricetus*) - issues and protocols. — In GODMAN, O. (ed.): Proceedings of the 9th Meeting of the International Hamster Workgroup, Bacharach: pp. 167-177. — Jb. Nassauisch. Ver. Naturkd.

KARASEVA, E.V. 1962: A study of the peculiarities of territory utilization by the hamster in the Altai territory carried out with the use of labelling. — Zoologiceskij zurnal **41**: 275-285.

KAYSER, A. & STUBBE, M. 2002: Untersuchungen zum Einfluss unterschiedlicher Bewirtschaftung auf den Feldhamster *Cricetus cricetus* (L.) einer Leit- und Charakterart der Magdeburger Börde. — Halle: Ministerium für Raumordnung, Landwirtschaft und Umwelt

KAYSER, A., STUBBE, M. & WEINHOLD, U. 2003: Mortality factors of the common hamster *Cricetus cricetus* at two sites in Germany. — Acta Theriol. **48**: 47-57.

KIRN, N. 2004: Ontogenese des Europäischen Feldhamsters (*Cricetus cricetus*) unter dem Einfluß verschiedener prä- und postnataler Photoperioden. — Doctoral Thesis, Institute of Zoology, University of Veterinary Medicine of Hannover.

KLEIN, D.C. 1972: Evidence for the placental transfer of 3H-actyl-melatonin. — Nature New Biol. **237**: 117-118.

KLEIN, D.C., MOORE, R.Y. & REPPERT, S.M. 1991: Suprachiasmatic nucleus - the mind's clock. — New York, Oxford: Oxford University Press.

KRSMANOVIC, L., MIKES, M., HABIJAN, V. & MIKES, B. 1984: Reproductive activity of *Cricetus cricetus* L. in Vojvodina-Yugoslavia. — Acta Zool. Fennica **171**: 173-174.

LEIRS, H. 2002: Conservation advices based on rodent pest biology: the case of the hamster. — In MERCELIS, S., KAYSER, A. & VERBEYLEN, G. (eds.): Proceedings of the 10th Meeting of the International Hamsterworkgroup, Tongeren: pp. 82-84. — Natuurpunt Studie.

LEMUTH, K. 2001: Ontogenese der circadianen und circannualen Melatoninrhythmik beim Europäischen Feldhamster (*Cricetus cricetus*). Student Thesis — Biological Institute, Department of Animal Physiology, University of Stuttgart.

MASSON-PÉVET, M., NAIMI, F., CANGUILHEM, B., SABOUREAU, M., BONN, D. & PÉVET, P. 1994: Are the annual reproductive and body weight rhythms in the male European hamster (*Cricetus cricetus*) dependent upon a photoperiodically entrained circannual clock? — J. Pineal. Res. **17**: 151-163.

MILLESI, E., PROSSINGER, H., DITTAMI, J.P. & FIEDER, M. 2001: Hibernation effects on memory in European ground squirrels (*Spermophilus citellus*). — J. Biol. Rhythms **16**: 264-271.

MONECKE, S. 2001: The two physiological identities of the European Hamster (*Cricetus cricetus* L.) - A race against the time of year. — In GODMAN, O. (ed.): Proceedings of the 9th Meeting of the International Hamster Workgroup, Bacharach: pp. 209-213. — Jb. Nassauisch. Ver. Naturkd.

MONECKE, S. & WOLLNIK, F. 2004: European hamsters (*Cricetus cricetus*) show a transient phase of insensitivity to long photoperiods after gonadal regression. — Biol. Reprod. **70**: 1438-1443.

MONECKE, S. & WOLLNIK, F. 2005: Seasonal variations in circadian rhythms coincide with a phase of sensitivity to short photoperiods in the European hamster, *Cricetus cricetus*. — J. Comp. Physiol. B **175**: 167-183.

MOORE, R.Y. 1996: Neural control of the pineal gland. — Behav. Brain Res. **73**: 125-130.

MOORE, R.Y. & LENN, N.J. 1972: A retinohypothalamic projection in the rat. — J. Comp. Neurol. **146**: 1-14.

MÜSKENS, G. 2005: Surviving of hamsters throughout the year. — Talk at the 13th Meeting of the International Hamster Workgroup, Illmitz/Vienna.

NECHAY, G. 2000: Status of hamsters Cricetus cricetus, Cricetus migratorius, Mesocricetus newtoni and other hamster species in Europe. — Strasbourg Cedex: Council of Europe.

NECHAY, G., HAMAR, M. & GRULICH, L. 1977: The Common Hamster (Cricetus cricetus [L.]); a Review. — EPPO Bulletin **7**: 255-276.

REITER, R.J. 1993: The melatonin rhythm: both a clock and a calendar. — Experientia **49**: 654-664.

REPPERT, S.M., CHEZ, R.A., ANDERSON, A. & KLEIN, D.C. 1979: Maternal-fetal transfer of melatonin in the non-human primate. — Pediatr. Res. **13**: 788-791.

REZNIK-SCHÜLLER, H., REZNIK, G. & MOHR, U. 1974: The European hamster (Cricetus cricetus L.) as an experimental animal: breeding methods and observations of their behaviour in the laboratory. — Z. Versuchstierkd. **16**: 48-58.

ROWE, S.A. & KENNAWAY, D.J. 2002: Melatonin in rat milk and the likelihood of its role in postnatal maternal entrainment of rhythms. — Am. J. Physiol. - Regul. Integr. Comp. Physiol. **282**.

SABOUREAU, M., MASSON-PÉVET, M., CANGUILHEM, B. & PÉVET, P. 1999: Circannual reproductive rhythm in the European hamster (Cricetus cricetus): Demonstration of the existence of an annual phase of sensitivity to short photoperiod. — J. Pineal Res. **26**: 9-16.

SCHMELZER, E. 2005: Aktivitätsmuster und Raumnutzung einer Feldhamsterpopulation (Cricetus cricetus) im urbanen Lebensraum. — Diploma Thesis, Department of Neurobiology and Behavioural Sciences, University of Vienna.

SELUGA, K., STUBBE, M. & MAMMEN, U. 1996: Zur Reproduktion des Feldhamsters (Cricetus cricetus L.) und zum Ansiedlungsverhalten der Jungtiere. [Reproduction of the common hamster (Cricetus cricetus L.) and the settlement of the young]. — Abh. Ber. Mus. Heineanum **3**: 129-142.

SIMONNEAUX, V. & RIBELAYGA, C. 2003: Generation of the melatonin endocrine message in mammals: a review of the complex regulation of melatonin synthesis by norepinephrine, peptides, and other pineal transmitters. — Pharmacol. Rev. **55**: 325-395.

SISK, C.L. & TUREK, F.W. 1987: Reproductive responsiveness to short photoperiod develops postnatally in male golden hamsters. — J. Androl. **8**: 91-96.

STEINLECHNER, S. 1992: Melatonin: an endocrine signal for the night length. — Verh. Dtsch. Zool. Ges. **85**: 217-229.

STETSON, M.H., ELLIOTT, J.A. & GOLDMAN, B.D. 1986: Maternal transfer of photoperiodic information influences the photoperiodic response of prepubertal djungarian hamsters (Phodopus sungorus). — Biol. Reprod. **34**: 664-669.

TIEGS, A. 2005: Untersuchung circadianer Parameter auf eine endogene circannuale Rhythmik beim Feldhamster (*Cricetus cricetus*). Student Thesis — Biological Institute, Department of Animal Physiology, University of Stuttgart.

VOHRALÍK, V. 1974: Biology of the reproduction of the common hamster, *Cricetus cricetus* (L.). — Vestnik Ceskoslovenské Spolecnosti Zoologicke **38**: 288-240.

VOHRALÍK, V. 1975: Postnatal development of the common hamster *Cricetus cricetus* (L.) in captivity. — Praha: Academia Nakladatelství Ceskoslovenské Akademie Ved.

WEAVER, D.R. & REPPERT, S.M. 1986: Maternal melatonin communicates daylength to the fetus in Djungarian hamsters. — Endocrinology **119**: 2861-2863.

ZUCKER, I. 2001: Circannual Rhythms - Mammals. — In TAKAHASHI, J.S., TUREK, F.W. & MOORE, R.Y. (eds.): Handbook of Behavioral Neurobiology, pp. 509-528. — New York, Boston, Dordrecht, London, Moscow: Kluwer Academic/Plenum Publishers.

ADDRESSES OF THE AUTHORS:

STEFANIE MONECKE & FRANZISKA WOLLNIK
Biol. Institute, Dept. of Animal Physiology
University of Stuttgart
Pfaffenwaldring 57
D-70550 Stuttgart, Germany

Present address: STEFANIE MONECKE
Institut des Neurosciences Cellulaires et Intégratives, Dept. Neurobiologie des Rythmes
UMR 7168/LC2 CNRS-Université L. Pasteur
5 Rue Blaise Pascal – 67000 Strasbourg - France
Stefanie.monecke@neurochem.u-strasbg.fr

Yearling male Common Hamsters and the trade-off between growth and reproduction

KARIN LEBL & EVA MILLESI

Abstract: Male Common Hamsters (*Cricetus cricetus*) continue structural growth after their first hibernation. Since *C. cricetus* are reproductively active after their first hibernation, yearling males are confronted with a trade-off in energy allocation for growth versus reproduction. To investigate how yearlings deal with this trade-off, we compared physiological parameters between two age groups, yearlings and older individuals. The hamsters were captured with Tomahawk live-traps baited with peanut butter. Body mass, testes width and moult were measured at capture. The animals were marked individually with PIT-tags and hair dye. At the beginning of the active season, yearlings were significantly lighter than the older individuals, but later they gained more mass, and at the end of the active season body mass was similar in both groups. Yearling males had a shorter reproductive period than older individuals and used the time before and thereafter for structural growth. Moult was less pronounced in yearlings during the mating season compared to older males. This could be an additional adaptation to allow energy allocation for growth. The results of this study demonstrate the trade-offs in energy allocation for different seasonal processes like reproduction, moult or preparation for hibernation and developmental factors like growth.

1 Introduction

Common Hamsters are, like many other hibernating mammals, not fully grown after their first hibernation (reviewed in ARMITAGE 1981). Therefore, in spring yearling males are smaller than older individuals. When males compete to gain access to oestrus females, smaller males are less likely to succeed in acquiring mates. Therefore, in many species, yearling males do not reproduce and use their first year for growth (DOWNHOWER and ARMITAGE 1981: *Marmota flaviventris*; SHERMAN and MORTON 1984: *Spermophilus beldingi*; MICHENER 1984: *Spermophilus armatus, Spermophilus beldingi, Spermophilus townsendii*). However, in some ground squirrel species one-year-old males are fully grown and reproductively active (MICHENER 1984: *Spermophilus parryi, Spermophilus richardsonii*). In *Spermophilus richardsonii* 80-90 % of the sexual mature individuals (≥1 year old) are yearlings, and there are only few older ones (MICHENER 1989). Because of the short lifespan in this species precociousness seems to be adaptive. In European ground squirrels (*Spermophilus citellus*) the timing of puberty seems to vary with population density. At high density, most males were immature after their first hibernation and used the following season for growth and fattening. But at low density, with fewer competitors and higher female availability, almost all yearling males were sexually mature, and reproduced successfully (MILLESI et al. 2004).

Usually male Common Hamsters start reproductive activity as yearlings, but sometimes they even can become reproductively active in the year of their birth (VOHRALIK 1974; NECHAY et al. 1977). Early maturation appears to be adaptive in species with short life spans, like *C. cricetus* because the chance to survive to the next breeding season is quite low (SAMOSH 1972).

The aim of this study was to compare physiological and developmental parameters of yearling and older male Common Hamsters during the active season to investigate the interactions between growth and reproduction.

2 Material and Methods

We investigated free-living male Common Hamsters in an urban area in the South of Vienna, Austria. The study site included a 5.6 ha park, in the vicinity of an apartment complex.

According to their daily activity patterns (SCHMELZER 2005) we captured the hamsters in the morning and in the evening hours from March until October 2004. We tried to capture each individual once to twice a week. We used Tomahawk live-traps baited with peanut butter. A cone-shaped cotton bag was used to handle the animals without anaesthesia (FRANCESCHINI 2002). Body mass and testes width were measured at capture. Moult was classified by a short pick on the fur in a standardized way. Depending on how much hair was pulled out we rated between (1) no, (2) few, (3) medium and (4) high stage of moult. The hamsters were individually marked with subcutaneously injected PIT-tags and with commercial hair dye.

We discriminated between three age classes: (1) juvenile – before their first hibernation; (2) yearling – after one hibernation period; (3) adult – after two or more hibernation periods. A definite age assignment for yearlings and adults was only possible for recaptured individuals. The age of individuals first captured in 2004 was estimated on the basis of their weight at spring emergence.

Three periods were distinguished during the active season. The posthibernation phase lasted from spring emergence (mid March) until the onset of the mating period. The mating phase lasted from 01.05.04 to 11.08.04 and was determined on the basis of calculated conception-dates (first observation of a litter minus 19d gestation and 19d for lactation in the burrow, BACKBIER et al. 1998; EIBL-EIBELSFELDT 1953). These calculations were supported by data on female reproduction in the study area over a 3-year period (FRANCESCHINI-ZINK & MILLESI 2008). From this calculated conception-dates we determined the span between percentile P_{10} and P_{90} as mating period. We divided this phase into an early and a late mating period (early mating phase: 01.05.2004–20.06.2004, late mating phase: 21. 06. 2004–11. 08. 2004). The mating period was followed by the prehibernation phase, which ended when the animals started to hibernate (early October).

3 Results

The course of body mass changes during the active season showed, that in spring (March–April) males could clearly be assigned to one of two weight classes (Fig. 1). The age class concurred with weight class in animals with known age. In this study, we labelled light individuals as "yearlings" and the heavy males as "adults" (discriminant analysis: Wilks' Lamda = 0.316, p < 0.001).

Both groups showed an increase in body mass in spring. Body mass was quite stable or rather decreased during the mating season. Mass increased again in late summer (Fig. 1). The weight of yearlings captured at the end of the active season was nearly undistinguishable from that of older males.

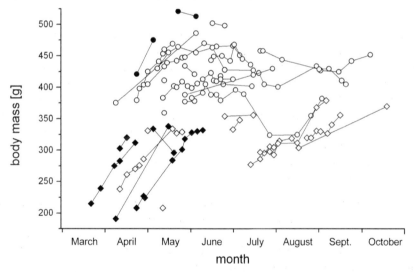

Fig. 1: Seasonal body mass changes in male hamsters. Yearlings: squares, adults: circles. Values of individuals were connected. Open symbols represent individuals of unknown age, animals with known age are marked by filled symbols.

During the posthibernation (March – April) and the mating phase the two age classes differed significantly in body mass, whereas during the prehibernation phase no significant differences were found (Table 1).

Table 1. Body mass in the two age classes during the three phases. Number of individuals (N), mean values (M), standard derivation (SD) and level of significance (p, Mann-Whitney-U-tests) are shown.

	posthibernation			mating			prehibernation		
	M	SD	N	M	SD	N	M	SD	N
yearling	279.4	49.89	9	311.1	38.89	11	344.4	33.91	4
adult	400.9	37.68	6	441.7	40.08	18	407.8	35.07	3
p		0.001			<0.001			0.157	

Yearlings showed a slight (but not significant) body mass increase from the posthibernation to the mating phase, as well as from the mating to the prehibernation phase (phase 1–phase 2: $Z=-1.481$, $p=0.138$, $N=9/11$; phase 2–phase 3: $Z = -1.4386$, $p = 0.151$, $N = 11/4$). In the prehibernation phase yearlings were heavier than during posthibernation ($Z = -2.006$, $p = 0.045$, $N = 9/4$). Adult males could increase their body mass from posthibernation to the mating phase ($Z = -2.067$, $p = 0.039$, $N = 6/18$), but showed no significant mass changes thereafter.

During posthibernation both groups showed a mass increase of about 2,9 g/d (Fig. 2). Mass changes in adult males differed significantly between the posthibernation phase and both parts of the mating period (phase 1–phase 2_1: $Z = -2.381$, $p = 0.017$, $N = 5/11$; phase 1–phase 2_2: $Z = -2.191$, $p = 0.028$, $N = 5/6$). In adults, body mass decreased during both early and late mating. In yearling males, body mass remained constant in the early mating phase. In contrast to older males, yearlings started to gain weight during the later part of the mating period. Mass increase rates differed significantly between the two age groups in this phase ($Z = -2.324$, $p = 0.020$, $N = 3/6$). In the prehibernation period both groups showed similar body mass increases (1.4 g/d).

Testes width changed during the active season (Fig. 3). Most individuals showed an increase in testes width in March and April. Testes size was quite stable during the mating period, and testes began to regress in late July. Testes regressed earlier in yearling than adult males. No significant difference in testes width could be found between the two groups throughout the season.

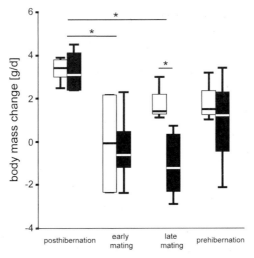

Fig. 2: Body mass changes per day of both age classes in the different phases. Positive values represent an increase, negative values a decrease in body mass. White boxes: yearlings, black boxes: adults. (N = 5/5, 2/11, 3/6, 3/3).

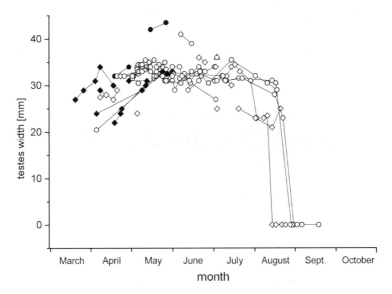

Fig. 3: Seasonal testes width changes in male hamsters. Yearlings: squares, adults: circles. Values of individuals are connected. Open symbols represent individuals of unknown age, animals with known age are marked by filled symbols. A testes size of zero indicates abdominal testes.

In the course of the active season a slight increase in moult could be found. Moult was less pronounced in yearlings during the mating season compared to adult males (Table 2). There was no difference in the other periods.

Table 2. Moult of yearlings and adult hamsters during three stages of the active season. Number of individuals (N), mean values (M), standard derivation (SD) and level of significance (p, Mann-Whitney-U-tests).

	posthibernation			mating			prehibernation		
	M	SD	N	M	SD	N	M	SD	N
yearling	0.64	0.732	7	0.840	0.633	11	1.643	1.085	3
adult	0.91	0.429	7	1.249	0.372	18	2.250	0.661	4
total	0.78	0.594	14	1.09	0.518	29	1.90	0.916	7
p		0.324			0.052			0.471	

Yearling and adult males differed in the timing of the active season (Fig. 4). Although yearling males emerged earlier from hibernation (end of March) than older males (mid-April), adult individuals completed testes development earlier. Yearlings had a shorter reproductive period, mainly because they ended reproduction earlier than older males. Yearling males started testes regression in late July, one month earlier than adults. After testes regression three of four individuals increased body mass. Both groups immerged into hibernation in late September. Therefore, after terminating reproductive activity, yearlings had twice as much time until the onset of hibernation than adults. In this period yearlings increased their body mass by 19.8 %.

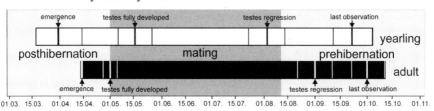

Fig. 4: Timing of the active season and reproductive activity in yearling and older male hamsters.

4 Discussion

Like other hibernating mammals, Common Hamsters have to reproduce, grow and prepare for hibernation within a limited time frame of about six months (NECHAY 2000). The mating period is therefore temporarily restricted and associated with high energetic costs (BRONSON & HEIDEMAN 1994). Because of these time and energy constraints, yearling males are confronted with a trade-off between growth and reproduction. Yearling males are smaller than older individuals in spring and have to invest energy in growth in order to reach the size of an adult male.

Our classification of the age classes "yearling" and "adult" on the basis of body mass at spring emergence was reasonable due to several facts. We knew the age of recaptured individuals of the previous year (2003). All individuals born before 2003 where at least two years old in this study. After hibernation the recaptured males with known age could be clearly assigned to one of two weight classes. It is improbable that some of the males we classified as yearlings on the basis of their body mass were weak or sick adults, because none of them looked haggard or showed signs of a disease. On the other hand, it is also unlikely that we accidentally labelled a yearling male as "adult", because in this case yearlings would have had increased their body mass during winter by more than 200 g.

Yearling males captured later in the season can also be clearly discriminated from early born juveniles (the first litters were born in the end of May), because juveniles are not able to increase their body mass that fast. Juvenile males in laboratory conditions weighed 180-200 g after 60d (MOHR et al. 1973, VOHRALIK 1975).

From spring emergence until May, both yearling and adult males gained mass. This increase in body mass was important for both groups to be prepared for the mating season. In the early mating period, body mass remained stable or decreased. During the late mating phase yearlings started to gain mass rapidly, whereas body mass in the adults continued to decrease. This is probably caused by the higher and longer lasting mating effort in adult males (CLUTTON-BROCK 1998; SCHNEIDER 2004). Similar patterns were found in other hibernating mammals (BIEBER 1998, MILLESI et al. 1998; SCHLUND et al. 2002; FIETZ et al. 2004). Consequently, body mass in yearling and older individuals did not differ at the onset of hibernation. A similar pattern was found under lab conditions, with body mass of yearling and older male Common Hamsters not differing at the onset of hibernation (VOHRALIK 1975).

All yearling males had descended testes in spring. Testes development in the hamsters continued after spring emergence, as the testes size increased during the posthibernation phase. Adult males had fully developed testes in early May, which concurred with the onset of the mating period. Although yearlings emerged earlier from hibernation than older males, testes development was complete two weeks later than in adults. Yearling males may have needed additional time for structural growth. Testes remained at maximal size for three months, from May to August in adult males. In Common Hamsters, testes development is controlled by an endogenous, circannual rhythm, synchronized by photoperiodic changes (MASSON-PÉVET et al. 1994, SABOUREAU et al. 1999). Common Hamsters in laboratory conditions started testes regression at a photoperiod of 15-15.5 h light per day (CANGUILHEM et al. 1988). In the study area, these photoperiodic conditions occurred in late July. Yearling males regressed their testes at this time, but adult males delayed testes regression until the end of August. This delay seemed to be adaptive because some females still became oestrous and mated in late summer (FRANCESCHINI-ZINK & MILLESI 2008). Female stimuli might modify the effects of photoperiodic signals. Yearling males probably reached their energetic limits earlier than older males and terminated sexual activity to avoid mass loss and gain additional time for growth and preparation for hibernation. This is supported by the fact that body mass in most males, independent of age, increased after testes regression.

Yearling males are confronted with a trade-off. They have to grow and gain weight to catch up and compete with older males. All yearling males were sexually active because delayed puberty would be unfavourable in a species with a relatively short lifespan (SAMOSH 1972). The first part of the mating period is probably the most important phase for male reproductive success. Juveniles born early in the season may have better chances to survive their first winter, because they have more time to grow and prepare for hibernation (ARMITAGE et al. 1976, SELUGA 1996). Moreover, in the studied population, early born litters were larger than later ones (TAUSCHER 2005). Oestrus occurred more synchronously during the first weeks of the mating phase (FRANCESCHINI-ZINK unpublished data). This lead to higher mate availability and better chances for smaller males to acquire a mate. In contrast to yearlings, adult males, which were less limited by body condition tried to remain sexually active as long as possible in order to fertilise oestrous females later in the season. As for most adult males the current mating period is probably the last one they should invest more in mating activity than yearlings, which have a higher chance for future reproduction.

In mammals the function of moult is to replace damaged or lost hair and to adapt to the different weather conditions. Moult was more pronounced in adult males during the mating season than in yearling males. This difference in moult intensity may be associated with higher energy allocation for growth in yearlings during this period compared to older males.

Although the potential for intrasexual competition was high in our study area, severe fights were rarely observed and males seemed to avoid each other. In most confrontations between males the smaller one was displaced by the larger one (ADLAßNIG 2005; own observations). The chance of yearlings to be successful in agonistic interactions is probably quite low because of their small body size.

It is the females who ultimately decide with which male they mate. Although the mating system of C. cricetus is often described as polygynous, females often mate with more than one male within one oestrus phase (PETZSCH 1936; ADLAßNIG 2005; own observation). Therefore it appears that Common Hamsters have a promiscuous mating system and there may be a chance for a yearling to mate with an oestrous female after larger males have left. This strategy may be more successful in the early mating phase when female oestrous occurs simultaneously, and older males invest less time in mate guarding than later in the season to be able to mate with more than one female.

It has been shown in many small mammal species that females prefer larger males (SCHWAGMEYER & BROWN 1983, FISCHER & LARA 1999, CLINCHY et al. 2004). GRULICH (1986) found, that heavier males acquired more copulations than males with low body mass. Furthermore, hamster breeding programmes have shown that females only accepted males which exceeded their own body mass by ~100 g (pers. comm. HAFFMANS S., Dutch Breeding Programme). This could mean that age and size predict high genetic quality of males (MANNING 1985, KOKKO & LINDSTRÖM 1996, BROOKS & KEMP 2001).

The results of this study indicate that yearling and adult males differ in the timing of moult and reproduction. Yearlings had a shorter reproductive period and used the time before mating and thereafter for growth. We suggest that intrasexual competition, the possible preference of females for larger males, and their shorter reproductive period cause yearling males to have lower reproductive success than older males.

5 Acknowledgements

The study was supported by the Austrian Science Fund (FWF, Project P16001/B06). We thank C. ADLAßNIG, C. FRANCESCHINI-ZINK, I. HOFFMANN, E. SCHMELZER, B. TAUSCHER and I. TSCHERNUTTER for their help in field work and data analysis.

6 References

ADLAßNIG, C. 2005: Das Fortpflanzungsverhalten des männlichen Feldhamsters (*Cricetus cricetus*). — Diploma Thesis, University of Vienna, Austria.

ARMITAGE, K.B. 1981: Sociality as a life-history tactic of ground squirrels. — Oecologia **48**: 36-49.

ARMITAGE, K.B., DOWNHOWER, J.F. & SVEDSEN, G.E. 1976: Seasonal changes in weights of marmots. — Am. Midl. Nat. **96**: 36-51.

BACKBIER, L.A.M., GUBBELS, E.J., SELUGA, K., WEIDLING, A., WEINHOLD, U. & ZIMMERMANN, W. 1998: Der Feldhamster, *Cricetus cricetus* (L., 1758) - Eine stark gefährdete Tierart. — 4. Tagungsband der Internationalen Arbeitsgruppe Feldhamster (Hrsg.), Limburg.

BIEBER, C. 1998: Population dynamics, sexual activity, and reproductive failure in the fat dormouse (*Myoxus glis*). — J. Zool. **244**: 223-229.

BROOKS, R. & KEMP, D.J. 2001: Can older males deliver the good genes? Trends Ecol. Evol. **16**: 308-313.

BRONSON, F.H. & HEIDEMAN, P.D. 1994: Seasonal regulation of reproduction in mammals. — In KNOBIL, E. & NIELL, D.J. (eds.): The Physiology of Reproduction, Vol. 2, pp. 541-583. — New York: Raven Press.

CANGUILHEM, B., VAULTIER, J.P., PEVET P., COUMAROS, G., MASSON-PEVET, M. & BENTZ, I. 1988: Photoperiodic regulation of body mass, food intake, hibernation, and reproduction in intact and castrated male European hamster. — J. Comp. Physiol. A **163**: 549-557.

CLINCHY, M., TAYLOR, A.C., ZANETTE, L.Y., KREBS, C.J. & JARMAN, P.J. 2004: Body size, age and paternity in common brushtail possums (*Trichosurus vulpecula*). — Mol. Ecol. **13**: 195-202.

CLUTTON-BROCK, T.H. 1998: Introduction: Studying reproductive costs. — Oikos **83**: 421-423.

DOWNHOWER, J.F. & ARMITAGE, K.B. 1981: Dispersal of yearling yellow-bellied marmots (*Marmota flaviventris*). — Anim. Behav. **29**: 1064-1069.

EIBL-EIBELSFELDT, I. 1953: Zur Ethologie des Hamsters (*Cricetus cricetus* L.). — Z. Tierpsychol. **10**: 204-254.

FIETZ, J, SCHLUND, W, DAUSMANN, K.H., REGELMANN, M. & HELDMAIER, G. 2004: Energetic constraints on sexual activity in the male edible dormouse (*Glis glis*). — Oecologia **138**: 202-209.

FISCHER, D.O. & LARA, M.C. 1999: Effects of body size and home range on access to mates and paternity in male bridled nailtail wallabies. — Anim. Behav. **58**: 121-130.

FRANCESCHINI, C. 2002: Der Feldhamster (*Cricetus cricetus*) in einer Wiener Wohnanlage. — Diploma Thesis, University of Vienna, Austria.

FRANCESCHINI-ZINK, C. & MILLESI, E. 2008: Reproductive performance in female common hamsters. — Zoology **111**: 76-83.

GRULICH, I. 1986: The reproduction of *Cricetus cricetus* (Rodentia) in Czechoslovakia — Acta Sc. Nat. Brno **20**: 1-56.

KOKKO, H. & LINDSTRÖM, J. 1996: Evolution of female preference for old mates. — Proc. R. Soc. Lond. B **263**: 1533-1538.

MANNING, J.T. 1985: Choosy females and correlates of male age. — J. Theor. Biol. **116**: 349-354.

MASSON-PÉVET, M., NAIMI, F., CANGUILHEM, B., SABOUREAU, M., BONN, D. & PEVET, P. 1994: Are the annual reproductive and body weight rhythms in the male European hamster (*Cricetus cricetus*) dependent upon a photoperiodically entrained circannual clock? — J. Pineal Res. **17**: 151-163.

MICHENER, G.R. 1984: Age, sex and species differences in the annual circles of ground-dwelling Sciurids: implications for sociality. — In MURIE, J.O. & MICHENER, G.R. (eds.): The biology of ground-dwelling squirrels, pp: 81-107. — University of Nebraska Press, Lincoln, USA.

MICHENER, G.R. 1989: Sexual differences in interyear survival and lifespan of Richardson's ground squirrels. — Can. J. Zool. **67**: 1827-1831.

MILLESI, E., HUBER, S., DITTAMI, J., HOFFMANN, I.E. & DAAN, S. 1998: Parameters of mating effort and success in male European squirrels, *Spermophilus citellus*. — Ethology **104**: 298-313.

MILLESI, E., HOFFMANN, I.E. & HUBER, S. 2004: Reproductive strategies of male European sousliks (*Spermophilus citellus*) at high and low population density. — Lutra **47**: 75-84.

MOHR, U., SCHULLER, H., REZNIK, G., ALTHOFF, J. & PAGE, N. 1973: Breeding of European hamsters. — Lab. Anim. Sci. **23**: 799-802.

NECHAY, G., HAMAR, M. & GRULICH, I. 1977: The Common Hamster (*Cricetus cricetus* L.) - A Review. — EPPO Bull **7**: 255-276.

NECHAY, G. 2000: Status of hamsters *Cricetus cricetus*, *Cricetus migratorius*, *Mesocriteus newtoni* and other hamster species in Europe. — Nature and environment **106**, Council of Europe Publishing, Strasbourg.

PETZSCH, H. 1936: Beiträge zur Biologie, insbesondere Fortpflanzungsbiologie des Hamsters. — Kleintier und Pelztier (Leipzig) **1**: 1-83.

SABOUREAU, M., MASSON-PÉVET, M., CANGUILHEM, B. & PÉVET, P. 1999: Circannual reproductive rhythm in the European hamster *Cricetus cricetus*: Demonstration of the existence of an annual phase of sensitivity of short photoperiod. — J. Pineal Res. **26**: 9-16.

SAMOSH, V.M. 1972: Growth and development of *Cricetus cricetus* L. — Zool. Rec. (Kiew) **4**: 86-89.

SCHLUND, W., SCHARFE, F. & GANZHORN, J. 2002: Long-term comparison of food availability and reproduction in the edible dormouse (*Glis glis*). — Z. Säugetierkde **67**: 219-232.

SCHMELZER, E. 2005: Aktivitätsmuster und Raumnutzung einer Feldhamsterpopulation (*Cricetus cricetus*) im urbanen Lebensraum. — Diploma Thesis, University of Vienna, Austria.

SCHNEIDER, J.E. 2004: Energy balance and reproduction. — Physiol. Behav. **81**: 289-317.

SCHWAGMEYER, P.L. & BROWN, C.H. 1983: Factors affecting male-male competition in thirteen-lined ground squirrels. — Behav. Ecol. Sociobiol. **13**:1-6.

SELUGA, K. 1996: Untersuchungen zur Bestandssituation und Ökologie des Feldhamsters, *Cricetus cricetus* L., 1758, in den östlichen Bundesländern Deutschlands. — Diploma Thesis, Martin-Luther-Universität Halle-Wittenberg, Germany.

SHERMAN, P.W. & MORTON, M.L. 1984: Demography of Belding's ground squirrels. — Ecology **65**: 1617-1628.

TAUSCHER, B. 2005: Entwicklungsprozesse bei juvenilen Feldhamstern (*Cricetus cricetus*). — Diploma Thesis, University of Vienna, Austria.

VOHRALIK, V. 1974: Biology of the reproduction of the common hamster, *Cricetus cricetus* (L.). — Vestnik Ceskoslovenske Spolecnosti Zoologicke **38**: 228-240.

VOHRALIK, V. 1975: Postnatal development of the common hamster *Cricetus cricetus* (L.) in captivity. Rozpravy Ceskoslov. — Akad. ved. Rada Matem. Prirod. Ved. **85**: 1-48.

K. Lebl & E. Millesi

Addresses of the authors:
Karin Lebl
Research Institute of Wildlife Ecology
Savoyenstraße 1
A-1160 Vienna, Austria
Karin.Lebl@vu-wien.ac.at

Eva Millesi
Institute of Zoology
Department of Behavioural Biology
University of Vienna
Althanstraße 14
A-1090 Vienna, Austria
Eva.millesi@univie.ac.at